College Student
Psychological Adjustment

College Student Psychological Adjustment

Theory, Methods, and Statistical Trends

Jonathan F. Mattanah

MOMENTUM PRESS
HEALTH

MOMENTUM PRESS, LLC, NEW YORK

College Student Psychological Adjustment: Theory, Methods, and Statistical Trends

Copyright © Momentum Press, LLC, 2016.

First published in 2016 by
Momentum Press, LLC
222 East 46th Street, New York, NY 10017
www.momentumpress.net

ISBN-13: 978-1-60650-725-4 (paperback)
ISBN-13: 978-1-60650-726-1 (e-book)

Momentum Press Psychology Collection

Cover and interior design by Exeter Premedia Services Private Ltd., Chennai, India

First edition: 2016

10 9 8 7 6 5 4 3 2 1

Printed in the United States of America.

Abstract

Young people are attending college at record levels in the United States, with over 65 percent of high school seniors in 2013 matriculating at a four-year college or university. Although many students thrive in college, others experience significant psychological and emotional difficulties, which can delay their pathway through college and even cause them to drop out. The current book provides an in-depth look at the psychological adjustment of college students, exploring theories and methods for studying the adjustment process and the latest trends in adjustment outcomes. Aimed at advanced undergraduates, graduate students, and scholars in the fields of psychology, human development, and higher education studies, this book reviews a wealth of classical and contemporary research on college student developmental processes and adjustment dynamics. Ultimately, readers will gain a fresh perspective on the challenges and opportunities of the college years and possible avenues for intervention to help those students who are struggling and at risk for failure.

Keywords

college student adjustment, depression, emerging adulthood, high risk behaviors, historically black colleges and university, loneliness, mental well-being, parent–student attachment relationships, predominantly white colleges, self-esteem, separation-individuation from parents

Contents

Preface

A Case Study

Maggie is a 22-year-old senior in college who works 30 hours a week in a paid internship related to her field of study in college and who has been accepted into a prestigious master's program. She also recently got married to her boyfriend who she has been dating for the past six years. Although life is good for Maggie right now and the future is bright, things have not always been this way for her in college. Her freshmen year was a real struggle. Maggie had a very hard time transitioning from high school to college, leaving her parents, to whom she was deeply attached, and separating from her serious boyfriend from high school. She did not get along well with her roommate and did not fit in well to the college environment, preferring to go home almost every weekend. Normally a straight A student and socially outgoing, Maggie began staying in her room, skipping class, and developed serious panic attacks. She felt depressed and slept much more than usual. Over the winter holidays, her father noticed that something was wrong and urged Maggie to get some help. During the spring semester, Maggie began seeing a counselor at the college counseling center and began taking antidepressant medication for her anxiety and depression. Things got a little easier, but she still felt depressed, she still wasn't getting along with her roommate, and she still missed her boyfriend terribly. Maggie remembers that it was not until her junior year in college that things began turning around for her. The pivotal event was taking a class with one professor who really took an interest in her and piqued her interest in her chosen field of study. After the class was over, this professor invited Maggie to help proctor this same class the next semester and eventually Maggie began working in the professor's lab and he became her mentor and thesis advisor, once she decided to become part of a competitive honors thesis program within her department. At this point, Maggie says, she began really applying herself in her classes, not just "floating" through—she also met fellow classmates who took their studies as seriously as she did.

Maggie's story reflects many themes common among the experience of students in college. Students frequently experience significant distress during the initial transition to college, related to leaving home and saying goodbye to important high school relationships. This distress can manifest itself in serious psychological symptoms, as it did in Maggie's case, requiring therapeutic intervention and medication. Furthermore, as in Maggie's case, close relationships are markers of success and failure. Initially, Maggie did not get along with her roommate and did not fit in well with the college environment—these relational failures increased her distress. Later on, Maggie met a professor who cared for and mentored her and she met like-minded students. These relationships nurtured her and Maggie thrived and succeeded, not just academically, but socially and emotionally as well.

Of course, in some ways, Maggie's case is atypical as well. Not many college students marry while in college these days and very few marry their high school sweethearts! And, not all students succeed as admirably as Maggie seems to have done, securing a lucrative internship as a college senior and matriculating into a prestigious master's program right after college. In addition to her high intelligence and outgoing personality, Maggie had the benefit of very supportive parents, a loving partner, and an invested mentor. Noting Maggie's advantages highlights how challenging the college transition can be even for talented and well-supported students. Unfortunately, other students are not so lucky and struggle even more dramatically with the transition to college.

The Scope of This Book

The current book provides an in-depth examination of the adjustment challenges associated with the college experience. In the first chapter, I explore current trends in college attendance, noting the rise in attendance among minority students, women, and students with disabilities. I also discuss some of the mental and physical health symptoms that students in college are experiencing these days, reviewing recent surveys that highlight the relatively high rates of these symptoms that students are reporting. This chapter lays the groundwork for the idea that college is a stressful experience and that we need to understand better how students are managing that stress more or less successfully.

Chapter 2 reviews classic and contemporary theories of college student development, highlighting the importance of this time period as one in which young people experience tremendous growth in their self-development. The theories of Chickering, Perry, Astin, Tinto, and Arnett are given special attention and for each theorist, I review their ideas about how college students negotiate an increasingly rich understanding of the needs of the self and other in mutually rewarding relationships. It is important to understand that all of this self-development is taking place in the context of the other stressors of being a successful college student in today's society.

Chapter 3 provides an in-depth examination of the empirical research that has been done on college student adjustment. First, I define and describe the concept of adjustment to college. Then, I review a large amount of research that has focused specifically on five domains of adjustment and explore factors that predict how students fare in terms of their adjustment to college. Finally, I review methodological issues related to studying the adjustment process, including examining different research designs and characteristics that modify outcomes for certain groups of students.

Finally, Chapter 4 of the book provides an overview of the adjustment process and adjustment challenges for students of color in the United States. Given their history of attending predominantly White institutions, where discriminatory practices are not uncommon, students of color have faced unique challenges in adjusting to these institutions. We review those challenges in Chapter 4 as well as discussing factors that predict success. We also discuss briefly some of the advantages for students of color who choose to attend historically Black colleges and universities.

Overall, this book provides a useful overview of theory and research on the college adjustment process of majority and minority students in the United States today. For those readers who wish to go in more depth on this topic, I also recommend my other book on this topic, *College Student Psychological Adjustment: Exploring Relational Dynamics that Predict Success*, which examines how students' relationships with their parents, faculty mentors, roommates, friends, and romantic partners affect their adjustment to college.

Acknowledgments

I am deeply indebted to many people for helping me bring this project to fruition. First, I acknowledge my two external reviewers, Laura Holt at Trinity College and Christa Schmidt at Towson University who meticulously read the manuscript, provided detailed feedback on content, formatting, and style, and who have been wonderful colleagues for the past three years, in our collaborative research on college student romantic relationships. In addition, I have gained immeasurably from my collaborative research endeavors and intellectual discussions with a number of colleagues, including Phil and Carolyn Cowan at University of California, Berkeley, Michael Pratt at Wilfrid Laurier University, Joanne Davila at Stonybrook University, and Frederick Lopez at the University of Houston.

I also acknowledge my students who have been collaborative research partners on a number of studies of college student adjustment over the years and to a few, in particular, who took the time to help read or write aspects of this book. First, Jennifer Daks read the entire manuscript and provided detailed feedback on each chapter. Also, I am grateful to Pauline Minnaar who cowrote Chapter 4 of this book. In addition to these two students, I have worked with countless other students over the years who serve as inspiration for many of the ideas I discuss in this book. I am grateful to them all.

I also extend my gratitude to the very helpful editorial staff at Momentum Press who helped shuttle this project to its completion. In particular, I acknowledge Shoshanna Goldberg who first approached me with the idea of writing this book and to Peggy Williams, Senior Editor at Momentum Press, who has been very helpful in answering all of my questions as I have prepared the final version of this manuscript.

Finally, I wish to acknowledge the love and support of my wife, Allyson Mattanah, who has stood by me throughout this and all of my scholarly efforts through the years and my children, Jeremy and Nadia, who are on the cusp of emerging adulthood. My wish for them is much the same as any parent's wish at this critical juncture: that they grow into mature and thoughtful emerging adults who increasingly rely on their friends and romantic partners while realizing their parents are always there for them when they need them! This book is for them.

CHAPTER 1

The Lives of College Students in the 21st Century

How Many People Are Going to College in the 21st Century?

Going to college seems ubiquitous in the landscape of American culture. It is an unspoken assumption of most Americans, of almost all classes, that their children will go to college after they graduate high school. It is a topic of consuming interest around the dinner table with most high school students. So, have you thought about college yet? Where are you going to apply? Have you taken the SATs yet? How many schools are you applying to? While these questions and dialogue is extremely common (at least among the circles I travel in, which is an important caveat), and often can be annoying to the high school teenager who just wants to get his (or her) homework done and be left alone, things have not always been this way and the reality is that college attendance is not quite as ubiquitous as many in the middle and upper classes in the United States assume it to be.

What do we actually know about college attendance in the United States? As it turns out, the U.S. government has been keeping fairly accurate statistics about this topic for at least the last three-quarters of a century. These data are collected by the National Center for Education Statistics (the NCES), an agency housed within the Federal Department of Education. Each year, the NCES releases their Digest of Educational Statistics, which provides a comprehensive look at the numbers of students attending school at every level from kindergarten through graduate school. The latest numbers (from the 2012 Digest report), suggests that 21 million students were attending college at a degree granting institution as of the 2011–2012 academic year. Of these students, 14 million of them are under the age of 25 and 7 million are 25 or older. These

14 million students represent 42 percent of all people aged 18 to 24 years old. So, although college attendance is not nearly as ubiquitous as some might imagine, it is still very much on the rise, and at the highest levels it has been since the NCES started collecting these data. By comparison, 36 percent of students aged 18 to 24 years were attending college in 2001 and 26 percent were attending college in the latter half of the 1960s (when the NCES began collecting systematic data on college attendance). The NCES estimates that college attendance has increased steadily at a rate of about 3.25 percent per year from 2000 through 2012 (NCES 2012, Digest of Educational Statistics Report).

Another way to look at these statistics is to ask the question, how many high school graduates matriculate at college immediately upon graduation. The numbers here are even higher, with 66 percent of high school seniors in 2012 immediately transitioning to college. About 40 percent of those students matriculate at a four-year institution with another 26 percent matriculating at a two-year college. Clearly, not all of those students stay at college after they matriculate and retention rates are a very important indicator of students' successful transition to college. According to the NCES, first-year retention rates are about 79 percent across all university types (they range from a low of 51 percent at private, for-profit institutions to a high of 95 percent at the most selective universities) and the four-to-six-year graduation rate is 59 percent (56 percent for males, 61 percent for females) (NCES Condition of Education Report 2014). Not only is college attendance up overall, but the demographic landscape of the college environment has changed quite dramatically in the past 40 years.

Who Is Attending College and How Has This Changed Over Time?

Rise of Women in College

The first notable change in college attendance over the last 70 years has been the dramatic rise of women attending college. In 1947, 29 percent of all college students were female. By 1970, that number had risen to 41.2 percent. According to the latest NCES data, currently 57 percent of all college students are female. Similarly, in 2012, 71 percent of graduating high school females immediately transitioned to college compared with 61 percent of graduating high school males. These trends are seen at the postgraduate level as well, with women increasingly attending graduate and professional schools. As one example, women made up 8 percent of attendees at medical school in 1966 compared with 47 percent of attendees as of 2013 (Association of American Medical Colleges' Report, *State of Women in Academic Medicine* 2014).

Rise of Ethnic Minorities Attending College

Secondly, the numbers of ethnic minority students in the United States attending college has changed quite dramatically from 1976 to 2011. In 2011, Caucasian students not of Hispanic origins made up 61 percent of all college students compared with 84 percent in 1976. Across that time period, Hispanic students have seen an increase of 10 percent, Asian American students 4 percent, and African American students 5 percent. These numbers are even more dramatic when looking at the immediate transition to college. In 1990, only 49 percent of African American high school graduates immediately transitioned to college compared with 62 percent in 2012. Hispanics have seen a similar rise in high schoolers immediately transitioning to college from 52 percent in 1990 to 69 percent as of 2012 (NCES 2012, Digest of Educational Statistics Report).

Unfortunately, six-year degree completion rates as of 2012 for African American (40 percent), Hispanic (51 percent), and Native American (40 percent) students are significantly lower than for Caucasian (62 percent) and Asian American (70 percent) students, suggesting that ethnic minority students who have faced significant historical prejudice in educational environments may experience more difficulties with the transition from high school to college and with successfully navigating the college environment (NCES Condition of Education Report 2014). This issue will be addressed in detail in Chapter 4.

Students with Disabilities Attending College

Notably, college students with disabilities have been attending college at much higher numbers since the passage of the American with Disabilities Act (ADA) in 1990. Currently, 11 percent of all college students have a disability that qualifies for services under the ADA (NCES 2012, Digest of Educational Statistics Report). According to a survey research completed by the Cooperative Institutional Research Program (CIRP) at UCLA, the most common disabilities reported by students entering college in 2010 (the first wave of students educated under the ADA law) were attention-deficit hyperactivity disorder (ADHD; 5 percent of students responding to the survey), other learning disabilities (2.9 percent), or a psychological disorder (3.8 percent). All other physical disabilities accounted for 4.5 percent of students responding to this survey (Pryor et al. 2010). The increasing numbers of college students with disabilities, most especially psychological and educational disabilities, has meant that college counseling centers and university academic support services are being taxed to provide additional services to these students, along with serving the wider population of students across campus. This is particularly problematic because we know that the broader population of college students are experiencing significant physical and mental health symptoms and engage in a variety of high risk behaviors.

College Students' Mental and Physical Health Problems

Although college students are generally young, healthy, and excited about this next stage in their life, a number of professionals working with college students have pointed out that students often experience significant stress and consequently develop notable health difficulties. One group particularly interested in this issue is the American College Health Association (ACHA). The ACHA conducts an annual survey of college students to examine trends in their physical and mental health. This survey, the National College Health Assessment (NCHA), has been completed every year for the past 14 years, starting in 2000. The latest survey was conducted on 66,887 students from 140 diverse campuses across all parts of the United States.

Physical Health of College Students

The results of the latest survey suggest that although 91 percent of college students describe their overall health as "good to excellent,"

9 percent feel that their health is poor. Moreover, 56.2 percent of students reported being diagnosed or treated for a health problem in the past year (ACHA-NCHA II Report 2014). Many of these health problems may be the result of a compromised immune system resulting from stress; additionally the presence of these health issues may cause increased stress for these students. Some of the most common problems requiring treatment included: allergies, back pain, bronchitis, irritable bowel syndrome, migraine headaches, sinus infections, strep throat, and urinary tract infections. The survey also asked students whether physical or mental health problems impacted their academic performance, including receiving a lower grade on an exam or in class, having to drop a class, or having their research or internship work disrupted by a health problem. Students reported that a wide range of health-related problems did impact their academic performance including, in order of importance: stress, anxiety; sleep difficulties; cold/flu/sore throat; depression; ADHD; and alcohol use (ACHA-NCHA II Report 2014).

Engagement in High Risk Behaviors

In addition to data on students' physical health, the NCHA survey also provides rich data on students' engagement in high risk behaviors. The survey focused particularly on three areas: (1) violence and abusive relationships; (2) substance use; and (3) sexual behaviors. Table 1.1 provides a summary of students' experience or engagement in these behaviors within a one-year time frame. Particularly noteworthy in these data is students' heavy use of substances—we know that substance use peaks during this time period and that students are at increased risk for becoming involved in a sexually abusive relationship as a result of their use of substances. These issues have caught the attention of the general public as evidenced by recent news reports of sexual assault on college campuses (Erdely 2014) and President Obama's "It's on Us" campaign to help raise awareness of sexual assault on college campuses (www.whitehouse.gov). In addition to high risk behaviors, college students also report significant levels of mental health symptoms and distress, data on which I turn to next.

Table 1.1 *Percent of students involved in abusive relationships or high risk behaviors in the past year (according to data from the 2014 NCHA survey)*

	Percentage of students experiencing this behavior in (%)	Percentage of males in (%)	Percentage of females in (%)
Violence or abusive relationships			
Physical fights	5.6	10.6	3.0
Sexual touching without consent	7.6	3.5	9.7
Sexual penetration attempt without consent	3.2	1.0	4.3
Sexually abusive intimate relationship	1.8	1.0	2.2
Emotionally abusive intimate relationship	9.4	6.3	10.9
Substance use			
Alcohol use—past 30 days	66	66.1	66.1
Had five or more drinks last time partied	33.6	43	28
Drive after drinking	12.1	15	11
Marijuana use—past 30 days	19.8	23.9	17.6
All other drug use—past 30 days	14.4	23.3	9.9
High risk sexual behaviors			
Had four or more partners in past year	10.8	13.4	9.4
Had anal sex in past 30 days	5.2	7.0	4.2
Did not use birth control	7.3	7	7

Source: ACHA (2014).

Mental Health of College Students

The NCHA survey referenced previously also asks students about their current mental health. As shown in Table 1.2, students report experiencing

Table 1.2 Student reports of mental health symptoms within a 12-month period

Mental health symptom	Total percentage of students reporting this symptom at any time in past 12 months in (%)	Total percentage of males in (%)	Total percentage of females in (%)
Felt things were hopeless	47.8	40.1	51.5
Felt exhausted (not from physical activity)	82.6	73.4	87.3
Felt overwhelmed by all you had to do	87.1	77.7	92
Felt very lonely	60.6	52.2	64.7
Felt very sad	63.2	52.8	68.4
Felt overwhelming anxiety	54.7	42.4	60.9
Felt overwhelming anger	38.5	35.3	40
Felt so depressed that it was difficult to function	33.2	28	35.6
Seriously considered suicide	8.6	7.9	8.8
Attempted suicide	1.4	1.3	1.4
Intentionally cut, burned, bruised, or otherwise injured yourself	6.9	4.6	8

Source: ACHA (2014).

significant feelings of sadness, hopelessness, loneliness, anger, and being overwhelmed at least some of the time in the past year. In addition, a number of students report having been diagnosed or treated for a mental health condition within the past 12 months. Overall, 23.1 percent of students surveyed said they had been treated for at least one mental health condition. Of these conditions, the most common were anxiety, depression, ADHD, and insomnia. In addition, 76 percent of the students reported having to deal with situations that they found "traumatic or very difficult to handle." Some of the most stressful situations for students were (1) intimate relationships, (2) family problems, (3) personal appearance, (4) sleep difficulties, and (5) finances. Sleep seems to be something that is seriously affected by students' stress levels as only 49.4 percent of students report getting enough sleep to feel rested more than 3 days per week and 42.6 percent of students said that daytime sleepiness is a significant problem for them.

Finally, and perhaps most alarming, is the number of students who consider or attempt suicide in a 12-month period. Suicide is the third leading cause of death among youths aged 18 to 24 years (after accidents and homicide) and the second leading cause of death among college students (after accidents) (Drum et al. 2009). Precise estimates of the number of students who consider, attempt, or complete suicide in a given year are hard to determine, as students are not generally willing to discuss these issues and universities do not collect these data systematically. Nonetheless, the best estimates across multiple studies suggest that as many as 18 percent of students have seriously considered suicide across their lifetime, 6 to 10 percent seriously consider suicide in a 12-month period, about 1 to 1.4 percent have made a serious suicide attempt in a given year, and roughly 1,000 students die by their own hands each year (ACHA-NCHA II Report 2014; Drum et al. 2009; Emory University Suicide Statistics n.d.). Suicide is a complex phenomenon, closely tied to other mental health problems, including most prominently mood disorders and substance use, but it is worth noting that suicidal ideation and attempts are often connected to the relational problems that students face in college. In one study, students reported that family, friend, and romantic relationship problems were the primary factors in leading to suicidal ideation (Drum et al. 2009).

Are mental health problems on the rise among college students? This is a controversial question, one that is being debated currently among professionals in higher education. In one set of studies, researchers were able to show that among students presenting themselves to counseling centers, the nature and severity of their mental health difficulties showed significant increases from 1991 to 1997 sample. These researchers noted three problem areas that seemed to be getting worse across these two cohorts of students: (1) depression as associated with romantic relationship difficulties; (2) adjustment problems with friends and roommates; and (3) academic concerns. No changes were noted in sexual issues or eating concerns (Erdur-Baker et al. 2006). By contrast, another study found no increasing severity in the presenting problems of college students attending counseling sessions across a six-year period from 1986 to 1992, although these researchers did find small increases in the number of students presenting at counseling centers as "extremely distressed" (Cornish et al. 2000).

In an attempt to bring clarity to this important question, a group of researchers based at Penn State University, called the Center for Collegiate Mental Health (CCMH), have begun to collect systematic data on the nature and severity of mental health problems across college campuses. The CCMH has released an annual report each year for the past five years in which they detail the results of their survey of college students attending counseling centers. The CCMH asked about a range of symptoms and mental health related behaviors, including attending counseling sessions, taking medication, suicidal thoughts and behaviors, aggression, exposure to traumatic events, and use of drugs and alcohol. Because the CCMH has been collecting these data systematically over the past four years, they were able to examine trends in mental health symptomology. Figure 1.1 provides a summary of those trends from 2008 to 2012 (CCMH *2013 Annual Report*). As can be seen in that figure, many mental health symptoms and behaviors have remained fairly constant during the past four years, whereas some behaviors are showing small but notable increases, including serious thoughts of suicide and exposure to traumatic events. These data provide some support for the idea that at least among students who are seeking counseling services, the nature and severity of their mental health problems may be getting worse.

Summary and Conclusions

Looking across the data I have presented in this chapter, there is reason for both optimism and pessimism when looking at the lives of college students in the 21st century. On the one hand, students are attending college at higher rates in the United States than ever before in history. Additionally, the pool of students attending higher education is increasingly diverse, including a much larger percentage of women and students of color than any time before in the history of the United States. Slightly over a majority of the students who begin college complete their degrees within 4 to 6 years and we know that college graduates are more likely to

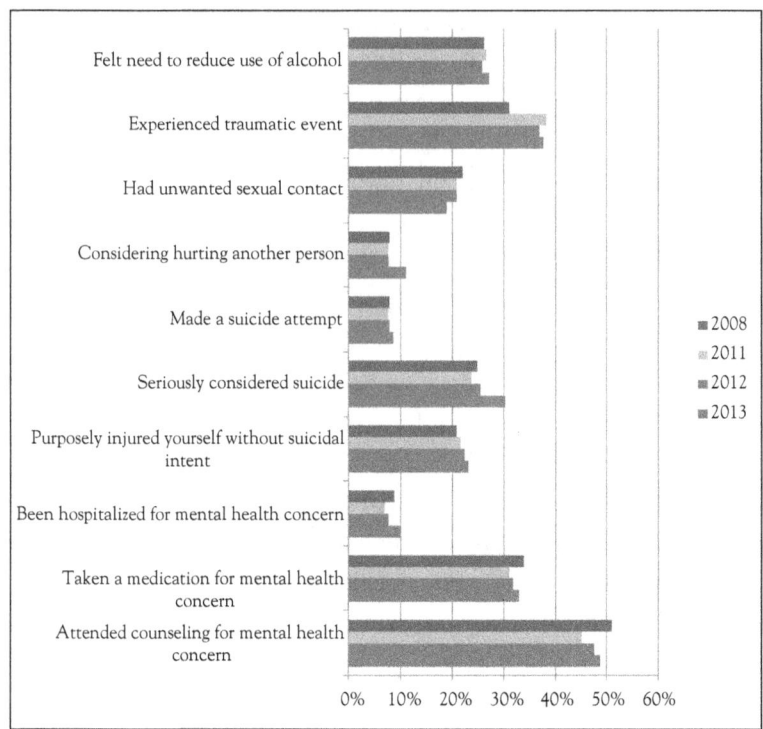

Figure 1.1 Five-year trends of student endorsement of mental health problems among students attending college counseling centers

Source: CCMH Annual Report (2013).

be employed and earn about twice the salary of those who do not attend college (NCES 2012, Digest of Educational Statistics Report).

On the other hand, students in college are stressed out! The majority of students report being overwhelmed at least some of the time and significant numbers of students report worrisome levels of sadness, anxiety, loneliness, and self-destructive behaviors. There are a number of reasons for these heightened levels of stress and anxiety, including serious financial concerns about how to pay for college (Eager et al. 2013). Increasingly, students are working part-time or full-time while in college, at the same time balancing their academic demands and trying to involve themselves in a wide variety of extracurricular activities. In addition to heightened stress levels, college students may appear more symptomatic than in the past because a wider range of students are attending college now, including students with significant mental disabilities, as discussed earlier.

Whether it is a changing demographic or increasing environmental stressors, it is imperative for researchers and professionals working in higher education to develop a deeper understanding of what allows students to succeed in college. By success, I am not simply talking about good academic performance and timely graduation, which are certainly important indicators of success, but I am also referring to social and emotional adjustment. Adjustment to college in the broadest sense will be the focus of the current monograph, in which I demonstrate the interwoven qualities of academic, social, and emotional success in college.

In the next chapter, I will examine a number of important theoretical perspectives that have been developed to understand how college students grow and develop in their understanding of themselves and in their relationship with other people as they move through their college lives. These theories lay the groundwork for understanding the college adjustment process.

CHAPTER 2

Theoretical Perspectives on College Student Development

Scholarly Approaches to Studying College Student Development

Having explained some of the challenges that college students face in Chapter 1, I turn my attention in this chapter to an exploration of the theoretical perspectives that have sought to explain the developmental pathways college students take in order to move from late adolescence into young adulthood. These theories are fascinating in their own right and they lay the foundation for understanding what is *normal* about college student development. We need to understand what holds true for college students, generally, before exploring the individual differences in why some college students succeed while others struggle with this developmental transition.

Before turning to specific theories of development, it is useful to understand who are the scholars studying college students. Three perspectives in particular have devoted considerable scholarly interest to the growth and development of college students and the issues they may struggle with as they make their way through college. The first are professors of higher education working at universities and institutes, who focus their attention on how college affects student growth and development as well as how institutions of higher education can design their programs to best promote student success. Notable scholars in this area include Arthur Chickering, Alexander Astin, Ernest Pascarella, William Perry, Vincent Tinto, and Patrick Terenzini. A number of these scholars have developed comprehensive theories of student development that will be reviewed in the following sections. A few prestigious journals that are devoted to the scholarly work of faculty in higher education research include the *Journal of Higher Education, Research on Higher Education*, and the *Journal of College Student Development*.

The second perspective is that of counseling psychology. Counseling psychologists have followed a model of promoting growth and supporting health as a way of preventing illness, rather than focusing on the amelioration of illness once it has developed, which tends to be emphasized by clinical psychologists. Most college counseling centers in the United States are staffed by counseling psychologists who not only see college students in need of mental health services but also aim to promote the mental and physical well-being of students across campus through campus-wide outreach programs and coordination with student organizations on campus. Counseling psychology faculty have devoted considerable attention to studying college student development. Notable scholars in the field of counseling psychology who have conducted studies on college student development include Maureen Kenny, Fred Lopez, Brent Mallinckrodt, Karen O'Brien, Kenneth Rice, and Meifen Wei. A number of journals are devoted to publishing scholarly efforts of counseling psychologists, including the *Journal of Counseling Psychology, The Counseling Psychologist*, and the *Journal of Counseling and Development*.

Finally, developmental psychologists have shown an increasing interest in studying college students as exemplars of a unique period of development. Early stage theories of development tended to focus primarily on childhood and adolescence, assuming that once the person turned 18 they were fully developed and ready to enter the world of adulthood. More recently, developmental psychologists have taken an interest in *lifespan* development, recognizing that individuals grow and develop across the lifespan and that development does not stop in adulthood. In 2000, Jeffrey Arnett, a developmental psychologist at Clark University with an interest in lifespan development, proposed that individuals in the 18 to 25 age-range, particularly in highly industrialized societies, are no longer adolescents and not yet adults (Arnett 2000). He suggested the term *emerging adulthood* as a new developmental period to capture the challenges and developmental changes associated with this time period. Of course, many emerging adults in the United States and other Western societies are in college during the formative years of emerging adulthood, making this theory very relevant to anyone interested in college students. Since Arnett made this proposal, there has been an explosion of research on emerging adult development, much of which has focused on college students. A new journal is devoted to publishing research in this area, called *Emerging Adulthood*, and a new society of scholars hold a biannual meeting to review research in this area, called *The Society for the Study of Emerging Adulthood*. I will review Arnett's interesting theory in greater detail, in the following sections, and suggest ways in which it may help capture many of the developmental challenges that college students face.

Higher Education Theories of College Student Development

Chickering's Seven Vectors of Development

As mentioned earlier, scholars of higher education have sought to understand the developmental trajectories of college students and examine ways in which college affects student development. One of the most influential theories in this regard was developed by Arthur Chickering. Chickering's background in lifespan developmental theory and higher education research led him to propose a theory that strongly emphasized identity development, much like the work of Erik Erikson (Pascarella and Terenzini 2005). Chickering has also been involved in higher education administration work himself and played a significant role in developing Empire State College in New York as its vice president of academic affairs from 1970 to 1977 (Chickering and Reisser 1993).

Most of us think of college as a place where students advance their education in hopes of attaining a specific degree that will prepare them for a particular career path. As Chickering explains in the preface to the second edition of his book, what made his theory unique within higher education research was his claim that college was about more than academic achievement and career readiness. On the contrary, Chickering claimed that college was about the growth of the total human being, a place where students explored their identity and developed increasingly mature relationships. As Chickering writes: "On these grounds, we argue for nothing less than human development, in all its complexity and orneriness, as the unifying purpose for higher education" (Chickering and Reisser 1993, xv). Chickering's claims are radical here, suggesting that college is an integral part of human development for those who choose to attend it. His claims are timely in the current political climate, in which emphasis is being placed on greater "accountability" in higher education outcomes. Those making these claims focus almost exclusively on achievement outcomes, such as grades, career placement, and graduation rates, while paying very little attention to the social and emotional consequences of a college education (Anderson 2015).

According to Chickering, the primary force that affects change in college students is disequilibrium. Disequilibrium means that students should experience challenges in college, encounters that radically shift their traditional ways of thinking, and allow them to move toward more complex ways of viewing the world, and to even reflect on their own thinking processes. In the first statement of his theory, Chickering emphasized autonomy as the major developmental task of young college students, being able to develop independence of thought and action. Chickering, following Erikson, proposed that autonomy precedes the development of intimacy in relationships. However, research conducted after the publication of the first edition of Chickering's book suggested that intimacy sometimes develops before autonomy, especially for women in college (Mather and Winston 1998; Straub 1987; Taub 1995). In his revised theory, Chickering modified his ideas suggesting instead a complex interplay between autonomy and relatedness and that college students may first learn autonomy in the context of relationships. As one student said in his project:

> I became less dependent on a relationship to feel good about myself or loved. The old way of thinking: I felt empty, lonely, and lost if I wasn't with a man. The new way of thinking: I don't need a man to be happy …. I knew I had really changed because I began to follow through with all my decisions. I was happy by myself. (Chickering and Reisser 1993, 27).

Chickering's most well-known contribution to higher education research is his proposal of seven vectors of development that take place during the college years. The vectors describe trajectories of movement that most college students will travel along at some point. Chickering proposed a sequence to his vectors, suggesting that movement along early vectors most likely occurs prior to movement in later vectors, although he did not see these vectors as stages of development per se. Figure 2.1 represents the seven vectors of development. In addition to the intellectual development traditionally associated with college life, Chickering's seven vectors emphasize emotional, interpersonal, and ethical development, all of which he considers essential to the maturation of the college student (Chickering and Reisser 1993).

In order to help validate Chickering's model of college student development, researchers have developed instruments to measure students' advances along the seven vectors across their years of college. Foremost among these measures is the *Student Developmental Task and Lifestyle Inventory* (SDTLI; Winston, Miller, and Prince 1987). Winston developed the

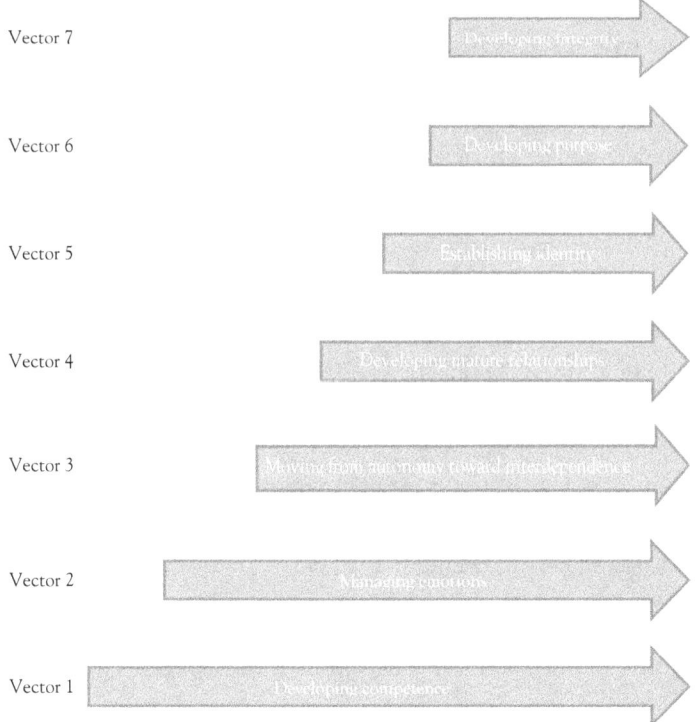

Figure 2.1 Chickering's seven vectors of development

Source: Chickering and Reisser (1993).

instrument to assess three of Chickering's seven vectors: (1) establishing purpose, (2) developing mature interpersonal relationships, and (3) moving through autonomy to interdependence, and found that these subscales were reliable and clearly distinguishable from each other (Winston 1990). Foubert et al. (2005) used the SDTLI to examine the sequence of vector development longitudinally from freshmen to sophomore to senior year. They found some support for Chickering's model in that developing mature interpersonal relationships (Vector 4) changed significantly from sophomore to senior year but not from freshmen to sophomore year, suggesting it is a later vector in the sequence. They also found changes in academic autonomy (a component of Vector 3) in each year of student's development. However, contrary to Chickering sequence, developing purpose (Vector 6) showed development from freshmen to sophomore to senior year, much earlier than Chickering would have predicted. Hence, the sequential nature of Chickering's vectors has received mixed support in the empirical research on his model.

Not only has Chickering articulated a vision for *what* develops in college, he has written extensively on those environmental factors that he believes foster such development. His emphasis is on colleges providing opportunities for students to engage actively with the university, to be provided learning experiences where they can reflect on the process of learning itself, and be given opportunities to contribute to the social and cultural life of their communities. Specifically, Chickering has emphasized (1) frequent opportunities for student–faculty interactions in multiple contexts, (2) a varied curricula that actively involves students in the learning process, (3) multiple opportunities for student friendships and communities based on meaningful subcultures, and (4) student involvement in the development of programs and services that serve their educational goals and are created in collaboration with faculty and staff (Pascarella and Terenzini 2005).

Perry's Model of Cognitive Development in College

A second major theory regarding growth and development in college comes from the work of William Perry, an educational psychologist who conducted an intensive, 15-year longitudinal study of undergraduate students at Harvard University in the 1950s and 1960s (Perry 1970, 1981). Although Perry's schema focuses specifically on cognitive growth, rather than encompassing multiple areas of development as in Chickering's model, it can be seen as a far-reaching theory that articulates the fundamental changes in thinking patterns and problem-solving that characterizes college student development. In fact, Perry's schema has been seen as foundational for most models of cognitive development in adulthood (Love and Guthrie 1999).

Based on open-ended interviews conducted with students during each year of their undergraduate career, Perry developed a nine-level scheme of cognitive growth. A simplified version of this scheme is shown in Figure 2.2. At the beginning of college, students view knowledge as something that exists "out there"; their task is to absorb and learn this

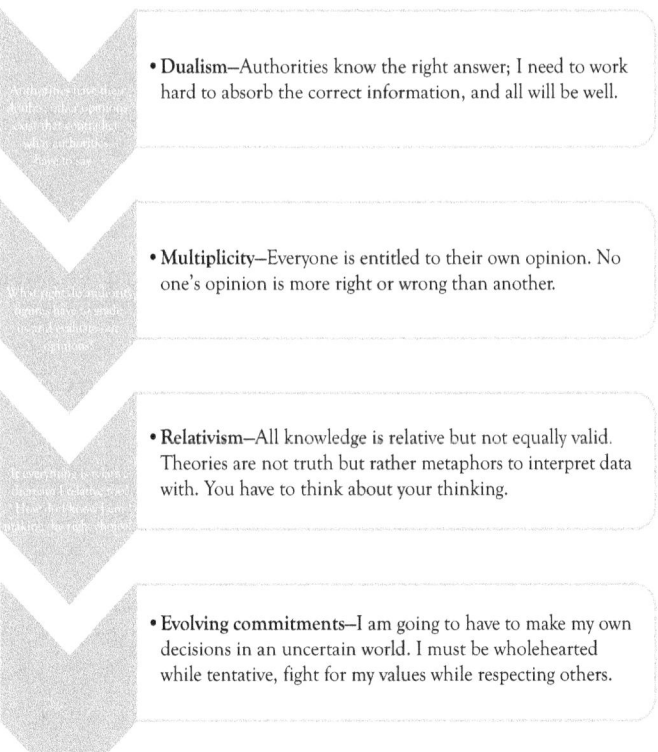

Figure 2.2 Perry's scheme of cognitive development in college

Source: Perry (1981).

information as fully as they can. In this early stage, students are dualistic in their thoughts, believing that authority figures know what is good, right, and true, and that other viewpoints are simply bad, wrong, or false.

Students first begin to shift away from this viewpoint when they encounter doubt and uncertainty. Fellow students may question the wisdom of a professor's statement or even a professor may express some doubt about the truth of what they know. The introduction of doubt disequilibrates the student's view of the world and causes the student to question whether there is a simple right and wrong answer to all issues. Dualism in thought gradually gives way to multiplicity, in which students believe that all opinions are equally valued and that there are no right or wrong answers. Everyone is entitled to their own opinion and no one opinion carries more weight than another one.

Multiplicity in thought is very freeing for the college student and is a common position held by many students in intellectual debates. However, students encounter a difficulty with this viewpoint when they begin to question why faculty are allowed to assign them grades, especially on subjective assignments such as critical analyses of literature, if everyone's opinions counts equally. Students' resistance to authority at this developmental stage makes them very suspicious of the whole intellectual enterprise in college, a crisis in thought that can lead to a temporary sense of disillusionment and bitterness. As one student put it: "This place is full of bull. They don't want anything really honest from you" (Perry 1981, 83).

This crisis pushes students to consider more deeply the criteria by which faculty are making evaluations and leads them to the realization that they are not being judged on the content of their ideas but rather on the manner in which they are making their arguments. Students gradually come to understand that faculty at universities are teaching them a *way* of thinking through problems, using logic and facts to draw tentative conclusions, open to revision with new information. This position is called relativism, in the sense that all knowledge and understanding is relative to the person articulating that knowledge, and needs to be evaluated based on the quality of the argument being proposed by the knower.

What is so important about this position is that students come to realize that they are contributors to the knowledge-creation process, not just absorbers of information known by others. In the final stage of Perry's model, students work toward committing themselves to things they value and believe, while recognizing that those commitments are tentative and that they need to respect the views of others as well, as long as those others can make articulate arguments for their positions. Perry's view of the mature college student thinker is very consistent with Chickering's perspective, wherein students develop an articulated, autonomous view of themselves and what they value, while deeply respecting and tolerating those who hold views different from their own.

Although foundational for other models of adult cognitive development, Perry's research has been criticized given its limited range of focus. Perry studied primarily White male students from elite backgrounds attending Harvard University in the 1950s and 1960s. This was a time period of much upheaval in society and students were very engaged in the process of challenging governmental institutions. Also, students were much more intentional and self-motivated in college at that time, when college attendance was more of a choice and less of a societal expectation, as it is today (Love and Guthrie 1999). Researchers have examined Perry's scheme in more diverse college settings and have found that although the general pattern of his developmental scheme fits a wide range of college students, there are differences in how far students advance along the positions of his scheme based on the type of university they attend (Love and Guthrie 1999). Additionally, women and men show differences in how they traverse the phases of Perry's scheme (Belenky et al. 1986). Nonetheless, most researchers of cognitive development in college emphasize the importance of developing a capacity for self-reflective judgment and consideration of multiple perspectives in constructing an argument, consistent with Perry's original scheme (see Baxter Magolda 1992 and King and Kitchener 2002 for more recent cognitive developmental models that have reworked and advanced Perry's original theory).

Astin's Theory of Student Involvement in College

Whereas Chickering and Perry's theories emphasized the *content* of development in college (*what* changes in college), a number of other higher education experts have studied *how* college brings about these changes. Foremost among these theorists is Alexander Astin, who developed the

Input-Environment-Outcome (I-E-O) model of college student development (Astin 1993, 1999). According to this model, students enter college bringing with them a host of background characteristics, their inputs, such as demographic variables, family background, and prior educational experiences. They then encounter the environment of college, which includes the full range of programs, people, and culture that students experience at college. Ultimately, as a result of these inputs and environmental experiences, students leave college with a set of outcomes, which includes the knowledge they have acquired, as well as a set of attitudes, beliefs, and behaviors that will drive their postcollege adjustment to the working world (Pascarella and Terenzini 2005).

For Astin, the key variable that predicts success in this process is student *involvement* in their college experiences. Astin claimed that involvement was the primary explanatory mechanism linking institutional programming to educational outcomes. Institutions of higher education place great emphasis on acquiring resources to enhance their institutional reputations, such as building large library collections, investing in expensive laboratory and research facilities, or hiring faculty with national reputations in their field. None of these resources do anything to predict successful student outcomes unless students are motivated to become involved in their institution, thereby taking advantage of these institutional resources (Astin 1999).

Astin viewed involvement in a quasi-psychoanalytic way, comparing it to Freud's ideas of cathexis. Cathexis, according to Freudian theory, refers to the investment of energy by a person into an object, idea, or another person. The more energy people invest into the "objects" of their world the more satisfied and well-adjusted they will be. Astin also saw involvement as similar to the psychological concept of motivation but prefers the term *involvement* to motivation because it is more behavioral and hence more easily observed and quantified. He claims that institutions can more easily develop programs to increase student involvement on campus than they can to increase student motivation.

In his own research on student involvement and academic outcomes, Astin found that three forms of involvement best predicted cognitive and affective development in college: (1) academic involvement, (2) involvement with faculty, and (3) involvement with student peer groups. Of these three, involvement with peer groups was the best single predictor of positive outcomes (Astin 1999). Astin also identified a number of forms of noninvolvement that negatively predicted academic and psychosocial outcomes. These included: (1) commuting to school, (2) living at home, (3) attending part-time, (4) being employed off-campus and full-time, and (5) watching television!

Finally, Astin conducted a fascinating analysis of the type of institutions that best foster student involvement and student success. He examined institutions in terms of how *research oriented* (defined as a high faculty publication rate, faculty spending a lot of time on research, and the institution showing a large commitment to research) versus how *student oriented* (faculty are interested and concerned with students and emphasize teaching and there are many opportunities for faculty–student interactions) they were. He found that the research-orientation of the school was associated with lower student satisfaction with faculty, development of public-speaking and interpersonal skills, and also lower GPA, likelihood of graduation, and overall satisfaction with college. The only positive correlation with research-orientation was students' GRE and LSAT scores. On the other hand, level of student-orientation of the school was correlated positively with institutional social activism, development of leadership skills, feeling a sense of community with faculty, and helping students to clarify their own values. Student-orientation also predicted greater degree attainment, graduation with honors, and growth in writing and critical thinking skills (Astin 1993, 1999). These findings are provocative and challenging, questioning many assumptions about the value of attending institutions based primarily on their "academic" reputations, which tend to be based on their level of research-orientation.

Astin's research and theory has had a major impact on institutional programming to enhance student involvement. In 1984, Astin's group released a report called *Involvement in Learning*, which provided a number of recommendations to enhance student involvement. Primary among these recommendations were to "front-load" resources in the first year to engender student involvement, greater use of active teaching modalities and the development of learning communities, improved academic advising and counseling, and more support for cocurricular activities to encourage involvement, especially among students who are traditionally least engaged in the campus community (e.g., commuter and part-time students). Many of these recommendations have been adopted by undergraduate institutions, especially programs such as Freshmen 101 courses, to help students adjust and get involved when they first come to campus. Also, many institutions have encouraged their faculty to develop more active, student-oriented, teaching modalities (Astin 1993).

Tinto's Theory of Student Integration into or Departure from College

Much like Astin's theory of student involvement as critical for student success in college, Vincent Tinto (1993) has developed an influential model to explain student decision making regarding staying versus leaving a particular university. Tinto emphasizes integration into the college community, a concept quite similar to Astin's notion of student involvement, but Tinto explores in greater depth the process of college integration.

For Tinto, integration is a process that takes time and is often painful and difficult for the student. Analogous to a "rite of passage," upon entering college, students must first separate from their past associations, both physically and socially. Tinto suggests that this separation is made easier by the support of old associates such as parents and the larger community. In the second phase, students transition to college, where students encounter "problems of adjustment whose resolution may spell the difference between persistence and departure" (Tinto 1993, 94). Tinto notes that during the transition phase students often feel weak, isolated, and normless. University transition programs, like the Freshmen 101 programs mentioned previously, seek to address the adjustment challenges associated with this transitional phase. In the final phase, incorporation occurs, where students feel socially and intellectually integrated into the college campus. Incorporation takes time and daily contact with fellow students, faculty, and other campus personnel. Students who fail to integrate into the college campus are most likely to leave after the first year of college, either transitioning to another university or dropping out of college altogether.

A simplified version of Tinto's longitudinal model of integration into and departure from college is shown in Figure 2.3. According to this

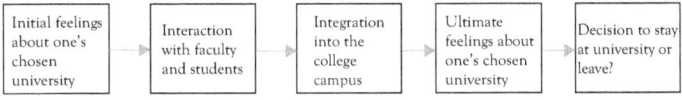

Figure 2.3 A simplified model of how students decide whether to stay at college or not

Source: Tinto (1993).

model, student background characteristics, such as family background and value systems, form the students' initial intentions and commitment to the goals and value of their chosen institution. Students then interact with the academic and social systems of the college community, which include formal activities such as classroom learning and membership in clubs and organizations, as well as informal activities, such as faculty–student interaction and informal peer interactions. These activities lead the student to experience some degree of academic and social integration, which then influences the students' goals and sense of commitment to the institution. Based on that new sense of commitment, students make their decision about whether to stay or leave the university.

The theories reviewed so far have highlighted some of the significant challenges that college students face during these important years of their development. More recently, developmental psychologists have suggested that young people of college age may actually constitute their own unique stage of development. I turn now to a more in-depth examination of that suggestion.

Arnett's Theory of Emerging Adulthood

In the May 2000 issue of the *American Psychologist* Jeffrey Arnett made a bold claim. He stated that young people aged 18 to 25 need to be designated their own developmental stage, distinct from adolescence and from adulthood properly (Arnett 2000, 2015). Arnett coined the term *emerging adulthood*, suggesting that youth of this age group were engaged in a prolonged, often arduous process of becoming an adult. In a sense, emerging adults exist in a state of moratorium between childhood and adulthood, not only with great freedom to explore themselves and their futures but also with much uncertainty and doubt about a range of issues.

Arnett based his claim on a number of factors. First, he pointed to dramatic changes in the demographic profile of individuals aged 18 to 25 in the past 50 years. In 1960, the average age of marriage was 21 for women and 23 for men, whereas in 2000 the average age was 26 for women and 28 for men. Accordingly, the average age for having children has moved about six to eight years later across this same time period. At the same time, as chronicled in Chapter 1, the number of individuals in this age group matriculating at college has risen dramatically, from roughly 35 percent of the population in 1960 to almost 70 percent of the population by 2010 (Arnett 2015). These changes reflect a radically different developmental trajectory for individuals during this time period. Rather than marrying and having children, traditional markers of entrance into adulthood, youths are going to college, figuring out what they want to do with their futures, and only then considering long-term relational commitments.

Along with these demographic shifts, Arnett pointed out that most individuals during this time period do not consider themselves to be a "full" adult, nor do they consider themselves adolescents. In Arnett's extensive surveys of youths aged 18 to 25, across the United States, he found that only about 40 percent said that they had reached adulthood by this time point (Arnett 2015). Arnett's participants felt that they had not yet mastered the three criteria they identified as marking someone as an adult: (1) accepting responsibility for one's self, (2) making independent decisions, and (3) becoming financially independent (Arnett 2000, 473). Yet, they were working toward each of these goals, which is very consistent with the developmental trajectories associated with college life described by higher education researcher reviewed earlier.

Arnett is not the first developmental psychologist to propose a new stage of lifespan development. Even the term *adolescence*, a word used in common parlance today, was popularized by the psychologist, G. Stanley Hall in the early part of the 20th century. Hall used the term to capture what he saw as a unique time period of sexual and physical maturity, accompanied by significant personality changes. However, historians have pointed out that Hall's construction of adolescence coincided with the changing landscape of youth life in late 19th-century and early 20th-century society, where a much larger percentage of the population was attending elementary and secondary schools and society was moving from an agricultural to a manufacturing economy (Kett 2003). The point is that developmental stages, as described by psychologists, need to always be understood within a particular economic and historical context.

For Arnett, four revolutions starting in the 1960s and continuing until today provided the context for the development of the concept of emerging adulthood (Arnett 2015). First, the technological revolution and rise of computerized manufacturing has meant that machines do most of the production line work in the United States today and that the United States has moved to a primarily service economy, including business, insurance, education, and health services. These service industry jobs require postsecondary education, explaining the dramatic rise in college attendance. Second, the invention of the contraceptive pill and the attendant sexual revolution of the 1960s have led to changes in the sexual morality of youths. It is much more acceptable, and actually normative, for young people aged 18 to 25 to engage in a series of sexual relationships before marriage. This change in sexual morality has taken the pressure of young people to get married at an early age in order to enter a sexual relationship. Third, the women's movement of the 1960s has led to important changes in how women view their development during this time period. Rather than marriage and motherhood as primary goals for young women, most women are now entering higher education institutions with the goal of developing a career before marriage. Finally, Arnett points to the youth movement of the 1960s and 1970s as a movement that "denigrated adulthood and exalted being, acting, and feeling young" (Arnett 2015, 6). He suggests that as a result of this movement young people do not see adulthood as something they are looking forward to anymore but more as a set of obligations that they plan to get to "eventually" but first they want to experience the pleasures of youth for as long as they can. The expression You Only Live Once (YOLO) captures this idea of wanting to have as many experiences as possible while you are young, before having to enter the somewhat domesticated, drab life of adulthood (Arnett 2015).

In an effort to elaborate his concept of emerging adulthood, Arnett has identified five features that he considers very characteristic (though not necessarily universal) of emerging adulthood (Arnett 2015). The first is *identity exploration*. It is during this time period that people really figure out who they are and what they want to do. Whereas Erikson had originally suggested identity exploration as the hallmark of adolescent development, Arnett claims that serious identity exploration occurs more so in emerging adulthood than in adolescence in 21st-century society. The second feature is *instability*. This time period is exceptionally unstable for many emerging adults. Emerging adults change their career plans, change relationships, even change where they live, more so during this time period than at any other period of the lifespan. The third feature is *self-focus*. As they have not yet entered into the obligations of adulthood but they have left many of the restrictions of their family-of-origins, emerging adults can be very self-focused in their decision-making process. They really do not need to consider others when making decisions about what they want to do or how they want to live their lives. Arnett is careful to distinguish between being self-focused and being selfish or self-centered. Although others have referred to youths today as dramatically selfish and increasingly narcissistic (Twenge 2013), Arnett argues that they are instead trying to become self-sufficient and don't want to rely on others in making their own decisions. The fourth feature is *feeling in-between*. This feature is almost definitional of this time period in the sense that emerging adults feel in-between childhood and adulthood. As mentioned previously, they see themselves as adult-like in some ways but not others and they view the process of becoming an adult as a gradual one. Feeling in-between can also contribute to emerging adults feeling unstable. The final feature Arnett calls *optimism/possibilities*. Emerging adults tend to feel optimistic about their futures and feel like they can easily change aspects of their life, even if things are not going well for them objectively. He suggests that because they have not yet experienced great disappointments, emerging adults maintain this optimistic outlook and see many possibilities for their future.

Arnett and others have done research to determine whether these features truly describe a unique developmental time period. In one set of studies, Arnett found that all five features were endorsed by individuals aged 18 to 25 to a greater extent than individuals in both older and younger age groups. In fact, 69 to 82 percent of individuals aged 18 to 25 endorsed items capturing these features as true for themselves (Arnett

2015). In these studies, socioeconomic status did not affect the rates of endorsement of these features. However, Arnett and others have pointed out that the features of emerging adulthood may not be cross-culturally consistent and that emerging adulthood may look quite different (or even not exist) in cultures that are not as industrialized as the United States and that place a much greater value on communal, interdependent relationships among family members. In one study conducted with Chinese participants, researchers found that caring for parents was a defining feature of adulthood for these participants, rather than accepting responsibility for oneself, a feature of adulthood more often endorsed by American samples (Zhong and Arnett 2014). As another example, northern Europeans tend to leave home around age 18 to 19, whereas southern Europeans tend to stay home with their parents until they enter marriage around age 30. Arnett suggests that this variation may make emerging adulthood a less unstable time for southern Europeans when compared with northerners (Arnett 2015). Research on emerging adulthood has really only just begun and researchers are actively exploring how well this concept captures the developmental changes that youth experience in countries around the world.

Clearly, the concept of emerging adulthood has great relevance to anyone interested in college student development and adjustment in the United States. Although not all emerging adults are college students (in fact, about 50 percent of individuals aged 18 to 25 are not in college in the United States, a population Arnett has referred to as the "forgotten half" [Arnett 2000, 476]), all traditional-aged college students are emerging adults, by Arnett's definition of the term. The defining features of emerging adulthood also seem consistent with theories of college student development. Identity exploration is central to Chickering's theory of college student development, whereas instability and feeling in-between seem consistent with Tinto's model of the transition to college as a rite of passage, involving significant stress and "normlessness." Arnett's theory of emerging adulthood adds to our understanding of college students by providing a framework for thinking about the internal experiences of individuals at this time period. Although exploring one's identity, feeling unstable, being self-focused, and feeling optimistic about the future are not unique to 18 to 22 year-old college students, they do describe many of the experiences of college students. I believe Arnett's characterization of emerging adulthood captures well the very mixed experiences of many college students who are stressed out, feel unstable and uncertain about their futures, while at the same time show high levels of optimism and maintain a positive spirit that things will work out well for them eventually.

Summary and Conclusions

In this chapter, I have reviewed a number of theories that examine the developmental challenges associated with college life. Each of these theories sees this time period as critical for the growth and maturation of the individual, as he or she moves out of the home, into a setting with an entirely new set of peers and adult mentors, and begins to outline the course of his or her adulthood. These theories detail the specific changes that need to take place for this transition to be successful, including developing a sense of self as a competent and autonomous person, negotiating intimate relationships with increasing recognition of how to balance the needs of self and others, learning how to express and manage difficult emotions such as anger, sadness, pride, and shame, and ultimately developing a value system that will help guide postcollege adult life.

Given the stress that students are experiencing with the transition to college, along with the developmental challenges associated with this time in their lives, it is no wonder that college student researchers are concerned with studying the adjustment process. In the following chapter, I examine in detail how scholars have defined and described the process of college student adjustment. I then provide an overview of a wide range of research studies that have explored what factors affect students' adjustment to college in five major domains of functioning covering academic, social, and emotional adjustment.

CHAPTER 3

Research on College Student Adjustment

The focus of Chapter 2 was on major theoretical models for understanding how students grow and develop during the college years. These theoretical perspectives are very important in conceptualizing the college adjustment process but they do not in and of themselves provide evidence for what actually happens when students go to college. In fact, most of the theories discussed in Chapter 2 grew out of the personal experiences of each theorist working closely with college students in counseling or advising roles. Personal experience is invaluable for developing a theoretical perspective but it does not in itself constitute scientific evidence. The current chapter moves from theory to research on the college adjustment process.

The focus of this chapter is on understanding what scientists know about college student adjustment. I will look at a number of issues in this chapter, including (1) how to define and conceptualize the process of adjustment; (2) the major domains of adjustment in college and what is known about factors that affect adjustment in each domain; (3) measurement strategies that have been used to assess college adjustment; and (4) individual characteristics that affect how some students adjust differently than others. But, first, I need to define what is meant by college adjustment.

What Is College Adjustment?

If you look up the word *adjust* in the dictionary, you will find the following meanings for the term: "1. to adapt or conform oneself (as to new conditions); 2. to achieve mental and behavioral balance between one's own needs and the demands of others" (www.merriam-webster.com/dictionary). These definitions imply that adjustment has to do with bringing something to a satisfactory state of being or with adapting one-self to a new situation. This captures quite well what I think is meant by college adjustment. For a student to adjust well to college they have to fit themselves to the college environment, meaning they have to actively seek to make their experience at college a suitable one that brings pleasure and feels healthy—mentally, physically, and psychosocially. Obviously, no student is going to adjust to college perfectly, all the time, but a significant degree of overall adjustment is needed for the experience to be a good one for most students.

When one considers the process of going to college, one can imagine a host of challenges that the young student faces. For many students, college is their first experience living away from home without their parents. They need to learn to take on responsibilities for managing their own finances, doing some basic household chores, such as laundry, dishes, and shopping for groceries. At the same time they have to adjust to an academic environment in which they are expected to function much more autonomously than in high school and navigate a largely new set of peer relationships, in which multiple temptations exist for distraction and engagement in risky behaviors (Mattanah, Hancock, and Brand 2004). To manage these challenges relatively well and experience an overall sense of success is another way to define positive college adjustment.

Five Domains of Functioning in College

In order to study the process of adjusting to college, researchers have focused on different aspects of students' functioning in college. This research has ranged very widely from examining academic performance to measuring loneliness and social adjustment to college to examining self-esteem and a sense of self-efficacy. In a recent meta-analysis on parental attachment and college student functional outcomes, my colleagues and I suggested a comprehensive model by which to organize the many functional outcomes that have been examined in the study of college adjustment (Mattanah, Lopez, and Govern 2011). The five domains we identified in this model were labeled: (1) academic achievement and academic competence; (2) interpersonal competencies and relational satisfaction; (3) stressful affects and high-risk behaviors; (4) self-worth and self-efficacy; and (5) developmental advances in autonomy, ego identity, separation–individuation, and career identity (Mattanah, Lopez, and Govern 2011, 584). Each of these domains is meant to be an overarching category within which are contained subcategories of functioning that are usually focused on within particular studies of college student functioning. So, for example, the domain of stressful affects and high-risk behaviors includes the construct of depression that is often studied when researchers are interested in examining how students are functioning in college in terms of their individual psychological functioning. In the sections that follow, I use the domain structure to review a number of studies that have been conducted on college student functioning in each domain and to examine predictors of functioning across these domains. I also spend some time discussing different kinds of measurements strategies that have been developed by researchers to examine these constructs. My intention is not to provide a comprehensive review of all research conducted in each area but rather to provide an introduction to research in these areas and give some idea of the scope of research being conducted on college student adjustment.

Domain 1: Academic Achievement and Academic Competence

Academic Performance as a Measure of Academic Competence

When considering academic competence, it seems obvious that one should focus on students' objective performance. In order to assess performance, researchers generally focus on college GPAs, which can be obtained from the students' academic record or can be self-reported by the student him or herself. Objective records are considered more valid although they can be hard to obtain, as most institutional review boards (IRBs) will require researchers to get specific permission from the student in order to access their academic record. Self-reported GPA is considered relatively valid as meta-analytic research has shown that self-reported GPA by college students correlates about 0.90 (range = 0.82–0.98) with GPAs obtained from school records (Kuncel, Credé, and Thomas 2005).

Although GPA is clearly an important indicator of academic success, it is by no means the only measure of academic adjustment and it is often poorly correlated with other measures of academic competence. Additionally, many variables that college student researchers are interested in looking at to predict student success, such as parent–student relationships, peer relationships, faculty–student interactions, and other individual and environmental variables, do not robustly predict college GPA (Halamandaris and Power 1999; Lopez 1997).

Given these challenges, researchers have sought to examine other indicators of academic competence that may help explain the link between relationship/interaction variables and academic performance. One early and influential study, conducted by Aspinwall and Taylor (1992), examined a large cohort of over 600 freshmen at the University of California, Los Angeles (UCLA) over a two-year period, in which data on self-esteem, optimism, and academic coping mechanisms were used to predict objective academic performance. These researchers found that more optimistic students, with greater self-esteem, were more likely to actively cope with academic problems and thereby enhance their performance in college over time. This kind of research is very important because it suggests that students with low self-esteem or high pessimism benefit from learning active coping strategies to deal with their academic difficulties in college.

Coping with Academic Challenges in College

A number of studies have followed the lead of Aspinwall and Taylor by examining the role of coping in students' academic functioning. Using the COPE Inventory (Carver, Schreier, and Weintraub 1989), which assesses engagement in active forms of problem-focused coping, researchers have found that while men and women do not differ in their use of coping, men are more influenced by a secure relationship with their fathers in terms of their willingness to turn to others for support, whereas women with good peer relationships are more likely to use active coping mechanisms (Greenberger and McLaughlin 1998).

In a related line of research, Simon Larose and his colleagues have developed an academic coping measure called the Test of Reactions and Adaptation in College (TRACS) (Larose and Bernier 2001; Larose and

Roy 1995). Across nine subscales, this measure assesses student's ability to prepare for exams, pay attention to their studies, and seek help from their teachers and peers when needed. Larose has found that the help-seeking scales are particularly important predictors of academic achievement in college and that in turn students who are less secure in their relationships with their parents and peers are less likely to seek help when needed. This makes intuitive sense and provides additional evidence that environmental factors (such as parent–student relationships) affect students' academic achievement through academic coping mechanisms.

The Importance of Study Skills

Finally, researchers have also examined study skills as one specific academic behavior that predicts academic achievement. One study used the Study Skills Questionnaire to ask students about a range of positive and negative studying behaviors (such as "reading the textbooks assigned for class" [positive behavior] and "watching TV or listening to music while studying" [negative behavior]; Norvilitis and Reid 2012). These authors found that the quality of one's study skills was the best predictor of higher GPAs, after controlling for students' inattention and hyperactive symptoms, personal motivation to succeed, and parental encouragement of intellectual activities.

Domain 2: Interpersonal Competencies and Relational Satisfaction

The second domain assesses the quality of students' social functioning in college. In particular, researchers are interested in the students' relationships with their peers, romantic partners, professors and other higher education personnel, and parents. Even though many students no longer live with their parents, parent–student relationships can have a significant impact on the social functioning of students while at college. There are three major ways in which researchers have assessed social functioning. First, they have looked at students' level of social support, referred to as the sense in which students feel adequately supported by their social networks. Second, research has examined the quality of students' intimate relationships and their reports of satisfaction in those intimate relationships. Finally, research has considered the experience of loneliness as an important indicator of a lack of adequate social functioning in college. I review research findings for each of these constructs in the sections that follow.

Social Support Processes in College

Social support has been studied extensively by psychologists as well as by many other behavioral scientists and has been shown to be extremely important to one's health and well-being. A sense of adequate social support predicts longevity and reduces health problems among older adults (Uchino 2004). Importantly, researchers have emphasized that social support is defined by the adequacy of one's support network in providing the support one needs rather than by its size. Hence, subjective self-report measures of social support are best used to capture this felt sense of adequacy.

One widely used measure of social support with college students, called the Social Provisions Scale (SPS; Cutrona 1989), targets students' relationships with peers, parents, and romantic partners. Cutrona and her colleagues found that social support predicted institutionally reported college GPAs, after controlling for the students' level of academic performance (ACT scores) in high school. She found that social support from parents was more important than support from peers or romantic partners in predicting academic performance. Students who felt adequately supported by their parents, particularly felt reassured of their worth, were less interpersonally anxious, had greater academic self-efficacy (meaning they felt more confident that they could do well in their classes), and these factors, in turn, predicted greater academic performance (Cutrona et al. 1994). This study adds to the studies I reviewed earlier in showing that environmental factors can predict academic performance particularly by enhancing students' sense of social well-being, which then enhances their academic coping and self-efficacy beliefs.

In studying social support processes, researchers have drawn a distinction between instrumental and emotional support (Larose, Guay, and Boivin 2002). Instrumental support refers to the provision of tangible forms of assistance and perhaps encouragements to help the student become more socially involved, whereas emotional support refers to listening and supporting the student when he or she is distressed. Larose, Guay, and Boivin (2002) found that emotional support in particular predicted less loneliness among students.

Finally, social support processes have also been studied in the context of students of color adjusting to campuses that contain primarily Caucasian students. Hinderlie and Kenny (2002) studied 186 Black students who were at six different predominantly White campuses. They measured social support from: (1) close friends, (2) student organizations, (3) instructors, and (4) parents. Hinderlie and Kenny (2002) found that all three categories of on-campus social support (i.e., friends, student organizations, and instructors) significantly predicted greater academic, social, and personal–emotional adjustment, and greater institutional attachment. Beyond the significant effects of on-campus support, social support from parents also predicted greater academic and personal adjustment but not social adjustment or attachment to the institution. These results make sense when considering that on-campus social support is probably more important to the students' sense of being integrated socially to the campus, whereas parents continue to support African American students' academic and personal adjustment.

Quality of Intimate Relationships

A second important dimension of social functioning in college focuses on the quality of students' intimate relationships. We know that students' capacity to form satisfying intimate relationships with romantic partners is an important predictor of students' mental and physical well-being in college (Braithwaite, Delevi, and Fincham 2010). Given this finding, researchers have spent some time examining what makes it difficult for students to form intimate relationships. One measure particularly well suited to examine intimacy difficulties is called the Inventory of Interpersonal Problems (IIP; Horowitz et al. 1988), which contains a "Hard to Be Intimate" Scale and a "Too Controlling" Scale. Mothersead, Kivlighan, and Wynkoop (1998) found that students who came from families with a history of alcoholism had greater intimacy difficulties, especially if there were high levels of family dysfunction present while the student was at college. In a similar study, Ensign, Scherman, and Clark (1998) found that a history of divorce and marital conflict were associated with less intimacy among students' romantic relationships. Clearly, family dynamics can play an important role in how students navigate their romantic relationships at college.

Loneliness: Measurement and Key Findings with College Students

Loneliness refers to the subjective experience of having "too few" social relations. Feelings of loneliness have been linked with alcoholism, delinquent behavior, suicide, physical illness, and overutilization of the health care system across the lifespan (Ponzetti 1990; Russell, Peplau, and Cutrona 1980). Students may experience high levels of loneliness as they transition away from their social networks in high school and attempt to integrate into new social environments in college.

The study of loneliness has been aided by the development of a psychometrically sound instrument that has been used widely in college student research. The UCLA Loneliness Scale contains 20 items that indirectly ask about the experience of loneliness (the word "lonely" is not used in any of the items). Ten of the items ask about negative experiences such as "There is no one I can turn to" or "I am unhappy being so withdrawn," whereas the other 10 items ask about positive experiences, suggesting a lack of loneliness, such as "I feel part of a group of friends" or "There are people who really understand me" (Russell, Peplau, and Cutrona 1980). The scale has been shown to be internally consistent and reliable across time. In terms of validity, Russell and colleagues have shown that scores on the UCLA Loneliness Scale are correlated with feeling "abandoned, depressed, empty, hopeless, isolated, and self-enclosed and with not feeling sociable or satisfied" (Russell et al. 475).

A number of studies have shown that loneliness has important implications for college adjustment. First, men tend to experience higher levels of loneliness than women overall and men who are lonely have lower college GPAs (Booth 1983; Ponzetti and Cate 1981). On the other hand, women who experience loneliness demonstrate greater physiological reactivity to stress, as measured by greater pulse pressure reactivity (divergence between systolic and diastolic blood pressure). In young college students, pulse pressure reactivity is a better predictor of long-term cardiovascular disease than systolic blood pressure alone, suggesting that lonely college women may be at greater risk for cardiovascular disease over time (O'Donovan and Hughes 2007). Interestingly, the women who reported greater social support, while also reporting medium-to-high levels of loneliness, showed significantly less pulse pressure reactivity. These results imply that social support can buffer the negative effects of loneliness on college adjustment. Given the importance of loneliness in college adjustment, researchers have begun to investigate whether peer-led social support interventions can ameliorate loneliness levels for young college students (Mattanah et al. 2010).

Domain 3: Stressful Affects and High-Risk Behaviors

A third domain of functioning that has been studied among college students examines psychological symptoms of distress and engagement in risky behaviors that can jeopardize one's physical or mental well-being (such as problematic drinking or promiscuous sexual behavior). As I wrote in Chapter 1, a surprisingly large number of college students struggle with depression, anxiety, eating problems, substance abuse, and other mental health difficulties while at college. For this reason, researchers have been focused on trying to develop instruments that accurately assess these kinds of problems with college-aged populations so as to better understand the nature and causes of such mental health difficulties and develop appropriate treatments for them. I begin this section by reviewing a very important new instrument that has been developed specifically to assess a range of mental health difficulties within college students. After that, I review research that has been conducted on students' struggles with a number of specific mental health problems and high-risk behaviors.

The Counseling Center Assessment of Psychological Symptoms (CCAPS-62)

In the past 10 years, a large group of counseling centers across the country have organized into a consortium in an attempt to study more systematically the mental health of college students. I referred to this group in Chapter 1 as the Center for Collegiate Mental Health (CCMH; based at Penn State University). This group of researchers and college counselors has argued that no specific instrument exists that measures a range of mental health difficulties specific to college students. Most instruments that do exist examine just one domain of functioning and they may not be geared specifically to college students. Hence, one of the first major goals of this group has been to develop a multidimensional instrument focused specifically on the mental health difficulties of college students.

This instrument is called the *Counseling Center Assessment of Psychological Symptoms* (CCAPS). It currently exists in 62- and 34-item versions that can be used to reliably and validly assess mental health difficulties across eight dimensions (Locke et al. 2011; McAleavey et al. 2012). Locke and his colleagues structured the instrument to target areas of challenge specific to college students (see Figure 3.1 for a list of the eight areas of concern targeted by this instrument). As they have access to over 90 counseling centers across the country, the CCMH group has been able to gather

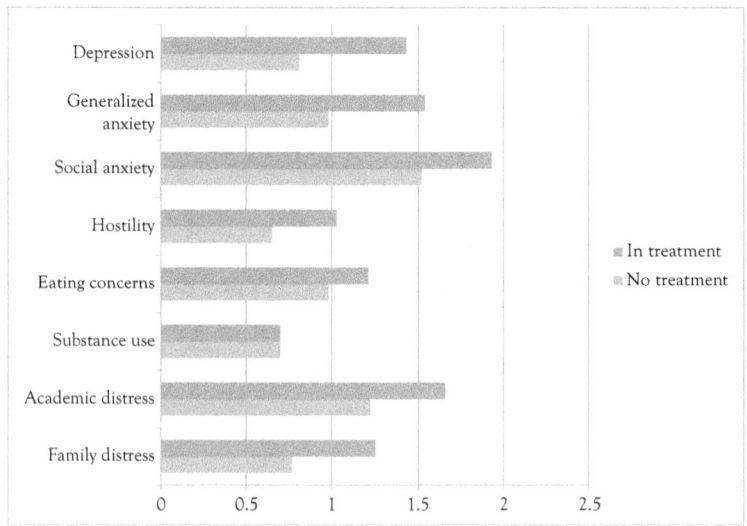

Figure 3.1 Symptomatic distress (on the CCAPS-62) for college students in counseling versus those not seeking treatment

Source: Data from McAleavey et al. (2012).

excellent validity data for their instrument, showing that it is substantially correlated with other measures of mental health and that scores on the CCAPS change significantly from before to after treatment. Importantly, however, the CCAPS is designed to assess mental health distress for both counseling center clients and for the more general population of college students. Locke and his colleagues have shown that the instrument is valid in both populations although students in counseling treatment demonstrate greater symptomatic distress than students not seeking treatment (see Figure 3.1, which demonstrates that counseling center students score higher on all subscales except substance use, suggesting that this problem is quite widespread across a range of college students).

The CCAPS is clearly an important development for the field of mental health counseling with college students and the instrument is being used increasingly by counseling centers across the United States to gather an initial assessment of students' mental health difficulties. However, as acknowledged by the CCMH group itself, the instrument is limited in its use for other purposes, partially because it is quite lengthy and also because it does not measure strength or resiliency in students (only symptomatic distress). In many studies of college student functioning, researchers have used shorter, more specific measures that examine one specific area of distress. I turn now to research that has focused on a number of those specific areas of distress.

Depression

One of the most common mental health symptoms reported by college students is depression. In the most recent survey of college student health, conducted by the American College Health Association (ACHA), 35 percent of female students and 28 percent of male students said that they "felt so depressed it was difficult to function" at some time within the past year and 14 percent of female students and 8 percent of male students said they had been treated for an episode of depression within the past year (results of the ACHA survey as cited in Whisman and Richardson 2015). As a group, Asian American college students show the highest depression levels, followed by Latino/a, African American, and then Caucasian students (Whisman and Richardson 2015).

Given the importance of this problem, there are a number of survey instruments to reliably measure depression in college students. The most widely used of these is the Beck Depression Inventory (BDI-II), currently in its second version. The BDI-II is a 21-item inventory asking about sadness, loss of pleasure, thoughts of worthlessness, guilt, self-criticism and suicidality, and somatic symptoms of depression, such as sleeping and eating habit changes, all occurring within the past two weeks. Each item is scored on a 4-point scale (from 0 to 3) yielding a total score from 0 to 63. Beck and others suggest that individuals scoring between 0 and 13 show minimal depressive feelings; 14 to 19 show mild depression; 20 to 28 show moderate depression; and scores 29 and above indicate significant or severe depression. The BDI-II is internally consistent and shows good test–retest reliability (Beck, Steer, and Brown 1996).

Research on depression among college students has shown that a number of factors increase the likelihood that a student will experience depression during the transition to college and some factors help protect against such feelings. First off, the experience of bullying and cyberbullying increases depressive feelings among college students (Tennant et al. 2015). In addition, students exposed to stressful life experiences, most particularly those experiencing sexual assault, reported significantly increased depressive symptoms during the transition to college (Chang et al. 2015). On the other hand, social support from peers as well as parents, has been shown to be an important protective factor, leading students to report fewer depressive symptoms in the face of stressful life events (Brandy et al. 2015; Tennant et al. 2015).

Anxiety

Anxiety is another major symptom that can accompany the transition to college. In studying anxiety among college students, researchers have tended to focus social anxiety and generalized anxiety. Social anxiety refers to the experience of feeling uncomfortable in social situations, being concerned that others are judging you negatively, and feeling afraid that one may embarrass oneself in a social situation. Many students experience social anxiety in college, sometimes heightened during the initial transition. Generalized anxiety refers to excessive worry and heightened physiological arousal that accompanies an overconcern with day-to-day stress and fear of failure.

Research on social anxiety among college students has used the Interaction Anxiousness Scale (IAS; Leary 1983; Leary and Kowalski 1993), which measures the subjective sense of feeling socially anxious rather than the behavioral symptoms of avoidance of social situations (a sample item is: "I often feel nervous when talking to an attractive member of the opposite sex" [Leary 1983]). Social anxiety is higher among students who leave for college versus those who commute to college while still living with their parents (Larose and Boivin 1998).

Research on generalized anxiety, using the Penn State Worry Questionnaire (PSWQ; Meyer et al. 1990), has found that worry symptoms are higher among women than men and that feeling secure in one's relationship with one's parents predicts less worry among college students (Vivona 2000). If one is more interested in assessing the physiological symptoms that accompany worry and anxiety, the Wahler Physical Symptoms Inventory (WPSI; Wahler 1968) assesses four areas of physical symptoms among college students: (1) gastrointestinal; (2) pain; (3) general systemic; and (4) sensory-motor-type symptoms. Physical complaints on the WPSI have been shown to be associated with greater depression, anxiety, hostility (especially for men), and interpersonal sensitivity (Holmbeck and Wandrei 1993). Additionally, more physical symptoms are associated with greater enmeshment with parents and with less adaptability to change, among women transitioning to college (Holmbeck and Wandrei 1993).

Anger, Aggression, and High-Risk Behaviors

In addition to distressing symptoms such as anxiety and depression, researchers are concerned with examining college students' tendencies to act out in inappropriate ways that may be harmful to themselves or others. The Weinberger Adjustment Inventory (WAI; Weinberger and Schwartz 1990) assesses students' ability to restrain themselves versus acting aggressively toward others. Research has shown that college men with better peer relationships and more empathy shower greater self-restraint on the WAI (Liable, Carlo, and Roesch 2004). These researchers suggest that a more positive peer climate may help these students to restrain themselves from acting aggressively and also that college student males who show greater self-restraint probably are able to form better peer relationships in the first place.

In addition to aggressive tendencies, a major focus of research on "acting out" behavior among college students has focused on alcohol use and abuse. As noted in Chapter 1, emerging adulthood represents a peak period for engagement in alcohol use, partially as a type of experimentation and separation from parents but also as a result of the stress and uncertainty associated with this time of life (Arnett 2015). The assessment of college student alcohol use focuses on two issues, first the frequency and intensity of students' use of alcohol and second the problematic consequences associated with alcohol use. To assess frequency, students are asked to report on how often they drink and how often they are intoxicated across a specific period of time, usually in the past year (Molnar et al. 2010).

In assessing problematic consequences of alcohol use, researchers can use well-validated and thoughtful questionnaires that assess a range of possible consequences of alcohol use specific to college students, such as the Rutgers Alcohol Problems Inventory (White and Labouvie 1989). Research has shown the consequences of alcohol use that are most problematic for students are physical symptoms such as withdrawal and "feeling like they are going crazy" and interpersonal consequences, such as having fights with a friend or family member or having family members avoid them because of their alcohol use (Neal, Corbin, and Fromme 2006). Clearly, alcohol use is particularly problematic for students to the extent that it disrupts their ability to succeed in school and maintain their most important interpersonal relationships.

Finally, it is also important to consider the motivation for students to drink. One frequently used questionnaire for this purpose assesses four possible motivations for drinking: (1) Social ("to enjoy myself at a party"); (2) Enhancement ("it's exciting"); (3) Conformity ("so that others won't kid me about not drinking"); and (4) Coping ("to forget about problems") (Cooper 1994). Research has shown that students who drink to cope are at the highest risk for developing problematic patterns of drinking. Secure attachment to parents makes it less likely that students will drink to cope with their problems and less likely that they will develop problematic drinking patterns (McNally et al. 2003).

The final area of high-risk behavior examined by college student researchers is engagement in risky sexual behaviors, including unprotected sexual intercourse, having sex while under the influence of drugs or alcohol, and having multiple sexual partners. Research on risky sexual behaviors has shown that having multiple sexual partners is associated with greater drug use among college students and that both drug use and sexual promiscuity were predicted by poor attachment relations with parents (Walsh 1995).

In more recent years, the phenomenon of "hooking up" has received much attention among those researchers interested in college students. This pattern of sexual activity focuses on brief physically intimate encounters with others who are not identified as romantic partners and with whom one has no plans of developing a longer term romantic relationship. Hooking up is quite common among college students, where as many as 60 to 80 percent of students in confidential surveys report having had at least one hook up experience (Garcia et al. 2012). Although hook ups are common and may be indicative of changes in normative patterns of dating and romantic involvement among emerging adults in the 21st century, there is research to suggest that some students are quite distressed after hooking up and long for more serious relationships with their hook up partners (Owen, Fincham, and Moore 2011).

Domain 4: Self-Worth and Self-Efficacy

Domain 4 focuses on two interrelated issues that are of great importance to students' ability to succeed in college and feel good about themselves. The first issue is self-esteem or self-worth and has to do with the student's overall assessment of him or herself as worthy and valued. The second issue carries multiple names, including self-efficacy, internal locus of control, or attribution style, but in each case the issue being considered is whether the student feels confident that he or she can complete tasks as a result of his or her own abilities or efforts versus attributing successes to external, uncontrollable forces, such as good fortune or luck.

Self-esteem is measured in college students using a variety of self-report questionnaires. Perhaps the most common and straightforward measure used in hundreds of studies is called the Rosenberg Self-Esteem Scale (Rosenberg 1965), which includes 10 statements capturing an overall sense of self-worth or what is called "dispositional self-esteem." Some sample items of this measure are: "I feel like I have a number of good qualities" or "All in all, I am inclined to feel like I am a failure" (reverse-scored). If one wishes a bit more subtle assessment of self-esteem, the Coopersmith Self-Esteem Inventory (CSEI; Coopersmith 1967) includes 25 items that assess attitudes toward the self in four areas relevant to college students: (1) social, (2) academic, (3) family, and (4) personal experiences. Research has shown that students from divorced backgrounds tend to have lower self-esteem but that parental acceptance and encouragement of independence promotes self-esteem for students from both divorced and nondivorced backgrounds (McCormick and Kennedy 1994, 2000).

In addition to feeling generally good about themselves, students need to feel a sense of self-efficacy in order to succeed at the academic and interpersonal challenges of college life. One assessment of self-efficacy has focused on the construct of internal locus of control, defined as the attribution of one's accomplishments to one's own efforts rather than to forces outside oneself. In an important study predicting students' objectively assessed academic performance in college, Fass and Tubman (2002) showed that internal locus of control independently predicted higher grades, after controlling for the students' intellectual ability, self-perceived scholastic competence, and high school GPA. Interestingly, an internal locus of control was, in turn, predicted by more secure relationships with parents and peers. In a second study, students with greater self-efficacy reported greater social support from their peers and experienced a more caring relationship with their mothers while growing up (Mallinckrodt 1992). Taken together, these results suggest that self-efficacy and self-esteem develop in an atmosphere of supportive and autonomy encouraging relationships with parents early on, which then helps students feel greater confidence about succeeding at their course work in college and developing supportive relationships with peers.

Domain 5: Developmental Advances in Autonomy, Separation–Individuation, and Career Exploration and Commitment

The final domain of functioning that has been focused on by college student researchers examines how well students are able to navigate a number of key developmental challenges associated with the emergence into adulthood. Some of these challenges were identified in Chapter 2 and are associated closely with specific developmental theories of emerging adulthood. Others are specific to the purposes and goals of a college education. I will review research on three interwoven challenges: (1) autonomy and identity development, (2) separation and individuation from parents, and (3) career exploration and commitment.

Autonomy and Identity Development

Research on autonomy and identity development has been strongly influenced by Erik Erikson's ideas about the centrality of the identity crisis in adolescence (Erikson 1968). Based on Erikson's model, researchers have argued that teenagers and emerging adult college students need to develop autonomy in order to explore their identity. Autonomy is defined in this context as the ability to make one's own decisions, become self-reliant, and not need others as much for emotional or instrumental support (Taub 1995, 1997). Along with the growth of autonomy comes a willingness to explore one's identity and ultimately to make a commitment to a chosen identity (Marcia 1966). Based on the two processes of exploration and commitment, Marcia identified four identity statuses for young adults. Those who have explored their identity and made tentative commitments to personally chosen goals and values are called *identity achieved*. Those who are fully engaged in the process of exploration but have not yet made commitments are in *moratorium*, while those who have made premature commitments to goals while foreclosing the process of exploration are labeled *foreclosed*. Finally, those who have not actively explore their identity nor have made any commitments are labeled *identity diffused*.

Research on autonomy development among college students has relied frequently on the Iowa Developing Autonomy Inventory (Jackson and Hood 1985), which measures autonomy development in three key areas: (1) emotional independence from parents and peers (lack of reliance on others to feel good about oneself); (2) instrumental independence (being able to take care of daily tasks oneself such as money and time management); and (3) interdependence (developing mutually supportive relationships with peers that replace dependency on parents).

Research using the Iowa scales has shown that senior college women are higher in autonomy than freshmen women, especially in terms of time and money management, and emotional independence from peers (Taub 1997). Taub notes that these changes did not take place for these women until senior year, presumably when students are looking toward future life plans, which may suggest that the development of these forms of autonomy are not as highly prized for women as for men and hence develop later in the college years. Importantly, Taub showed that women did not report changes in their level of closeness to parents across this time period, which challenges the idea that college women need to give up their dependency on and support of their parents in order to advance their own autonomy. In another study, Schultheiss and Blustein (1994) showed that it was the combination of secure attachment to parents along with a lack of guilt about separating from them that predicted greater academic autonomy and developing a sense of purpose for both college women and men, suggesting that the development of autonomy may not require the relinquishing of emotional ties to parents for either gender.

In a seminal set of studies, Jane Kroger (1985; Kroger and Haslett 1988) examined the development of identity across three years of college, using Marcia's (1966) Ego Identity Status Interview, which allows for a classification of students into the four identity statuses described previously (i.e., identity achievement, moratorium, foreclosure, and diffusion). Her results suggest that identity status is quite stable from first to third year for those students who are either identity achieved and have a secure relationship with their parents or who are in foreclosure and have an insecure relationship with their parents. Contrarily, those in moratorium tended to move toward identity achievement across this time period, especially if they had more secure relationships with parents. These results provide strong support for the argument that the process of identity exploration (moving out of foreclosure or moving from moratorium to achievement) is aided by having a relationship with one's parents in which exploration is encouraged in the context of ongoing support. Students who feel insecure in their relationship with their parents tend to remain foreclosed in their identity exploration process, which ironically often involves a premature identification with one's parents' goals and values with little willingness to explore alternatives. Given the importance of parents to the identity development process of college students, researchers have devoted some attention specifically to the process of separating and individuating from parents, to which I turn next.

Separation and Individuation from Parents

Research on the process of separating and individuating from parents has been based largely on the work of Margaret Mahler, who studied how infants and toddlers develop a sense of themselves as separate from their caregivers (Mahler, Pine, and Bergman 1975). Mahler suggested that for infants to feel separate from their caregivers they must first engage in independent behavioral activity, which helps them learn about the physical separation between self and others and then need to develop a view of themselves as cognitively and psychologically a different person from their parents (i.e., individuated). This process is challenging and fraught with emotional danger, as the infant fears losing the parent's love if they separate from them or alternatively being engulfed by the parent, if they resist separation. Ultimately, the healthy child develops a balance between individuation and maintaining a connected relationship with the caregiver.

Based on the early work of Mahler, later theorists suggested that a second separation–individuation occurs during adolescence, where the adolescent once again has to go through a process of separating and individuating from the views and beliefs of his or her parents, hopefully while still maintaining a connected relationship (Blos 1979; see also Lopez and Gover 1993). Based on these ideas, Hoffman (1984) developed a measuring of separation and individuation from parents called the Psychological Separation Inventory (PSI), which has been used extensively in studies of college students. Hoffman argued that separation–individuation can be different from mothers versus fathers and hence developed separate scales for mothers and fathers. The instrument contains four subscales that assess different aspects of the separation–individuation process. First, the *Functional Independence* subscale assesses the adolescent's ability to make independent decisions without consulting with parents first. The *Attitudinal Independence* subscale assesses the adolescent's willingness to have beliefs that differ from his or her parents. The *Emotional Independence* subscale assesses the adolescent's lack of excessive emotional dependency on parents. Finally, the *Conflictual Independence* subscale assesses the adolescent's freedom from guilt, excessive anxiety, or anger regarding the process of separating from his or her parents.

The PSI has been used extensively in college student studies. Men tend to score higher than women on functional, emotional, and attitudinal independence and these forms of independence increase from freshmen to junior year of college (Lopez and Gover 1993). Although advances in all four dimensions of separation were originally thought to predict better adjustment for college students, results suggest that only the Conflictual Independence subscale is consistently associated with better adjustment for students. In some cases, functional, attitudinal, and emotional independence are actually associated with worse adjustment outcomes, especially if they are not accompanied by secure relationships with parents. Once again, these results suggest that separation–individuation from parents occurs best in the context of a supportive relationship, which can actually facilitate the process of separation and thereby predict better adjustment outcomes (Mattanah, Hancock, and Brand 2004).

Career Exploration and Commitment

One of the most important developmental tasks associated with college is students' ability and willingness to explore their career identity. Much like the identity formation process described previously, in order to pick an appropriate career for themselves, students need to be willing and able to explore a variety of career options and ultimately show commitment to a chosen career identity. Researchers have developed measures of each of these processes and have explored factors that predict students' success in accomplishing these developmental goals. In terms of career exploration, the Career Search Self-Efficacy Scale (CSES; Solberg, Good, and Nord 1993) assesses students' confidence in exploring their career options in four dimensions: (1) job exploration; (2) interviewing; (3) networking; and (4) personal exploration. Ryan, Solberg, and Brown (1996) found that family dysfunction was associated with less career search self-efficacy for women, whereas a secure relationship with mothers and fathers was associated with greater search self-efficacy for both men and women. Similarly, Lease and Dahlbeck (2009) found that secure attachment relationships with parents were predictors of greater career decision-making self-efficacy for men and women but that interestingly a style of authoritarian parenting was also predictive of greater career-decision-making self-efficacy, specifically for women.

These results highlight the importance of parental relationships in supporting the career search process for college students but suggest that parents may play quite an active role here, almost needing to push their students to engage in this process, especially in the case of women. These results are particularly important when considering that women have been shown to decrease their career aspirations over the years of college and indicate that they are willing to settle for more traditional, less prestigious careers in order to balance family and work considerations (O'Brien et al. 2000). Finally, research on career commitment has found that a balance of conflictual independence from and secure attachment to parents predicts greater commitment to career choices (Scott and Church 2001).

Methodological Paradigms for Studying College Student Adjustment

In the previous sections, I detailed the five domains of functioning that have been focused on within college student adjustment research. I turn now to a discussion of the advantages and disadvantages of the primary methodological paradigms used in college adjustment research. By far, the most common research design used in this area is the *cross-sectional* study, in which data on predictor (such as parent–student relationships, personality variables, or college environmental variables) and outcome variables (such as loneliness, depression, self-esteem, or career identity development) are collected at one point in time. The obvious advantages of cross-sectional research designs are ease of data collection and no loss of participants in the study. Since all participants are participating just one time, there is little chance that they will discontinue their participation in the study over time. The major disadvantage of cross-sectional research is that because data are collected all at once it is very hard to disentangle causal relationships. By providing a time separation between predictor and outcome variables, as is done in longitudinal research designs, the researcher can make causal interpretations with a somewhat higher degree of certainty, especially if they are able to control for initial levels of the outcome variable at Time 1 and look for change in those variables over time. Hence, longitudinal studies provide a more compelling design for making causal conclusions but risk loss of participants over time, due to a variety of circumstances. Longitudinal studies are less common in college student adjustment research, but they have been completed and often provide useful information not available through cross-sectional research. In the next section, I review four major longitudinal studies that have provided rich data on the adjustment process over time in college.

Longitudinal Studies of College Adjustment

The first two studies I review are examples of short-term longitudinal studies because they gathered data on college student adjustment over the first semester of college life. This kind of study is not uncommon since students are easier to track down over one semester and also the first semester of college is considered a time of great upheaval where students' functioning may be quite variable and important to study. Pritchard, Wilson, and Yamnitz (2007) collected data on 242 students who completed assessments of optimism, perfectionism, coping tactics, and self-esteem levels (their predictor variables) during the first week of orientation to college (Time 1) and completed assessments of their physical health (21 health symptoms including cold, flu, shortness of breath, etc.), alcohol use, and negative mood symptoms (depression and anxiety) both at Time 1 and Time 2, one week before the end of the semester. The first question these authors were interested in was whether students' health got worse during the transition to college, a result that has been demonstrated previously in students transitioning to medical and law school. Since they had gathered data on their outcome variables at both Time 1 and Time 2 they were able to address this question. They found that students did indeed report increased health symptoms over the first semester of college. Additionally, they reported an increase in alcohol consumption on the weekends and increasing negative mood symptoms over this time period. These results certainly support the contention that the first semester of college is stressful for college students and can lead to increased mental and physical health difficulties, at least in the short-term. In terms of predictor–outcome relations, Pritchard, Wilson, and Yamnitz (2007) showed that low self-esteem, perfectionism, pessimism, and coping through denial and self-criticism all predicted worsening health symptoms and more negative mood symptoms over the first semester of college. Using alcohol to cope with stress was the only significant predictor of greater intoxication over time.

Using a similar paradigm, Holt (2014) studied 204 college students at the beginning and end of the first semester of college. She gathered data on students' sense of having a secure attachment relationship with their parents at the beginning of the semester (the predictor variable) and then gathered data on their academic adjustment to college at the end of the semester. Holt reasoned that a more secure attachment relationship with parents would make students more willing to seek help when they needed it in their classes (having a history of secure attachment to parents allows students to feel that others are available and responsive to their needs) and that greater help-seeking would ultimately lead to better academic adjustment. Hence, she also gathered data on students' attitudes and intentions to seek help (the mediator variable) at the beginning of the semester. Figure 3.2 provides a graphical representation of Holt's model, demonstrating the pathways she hypothesized from secure attachment

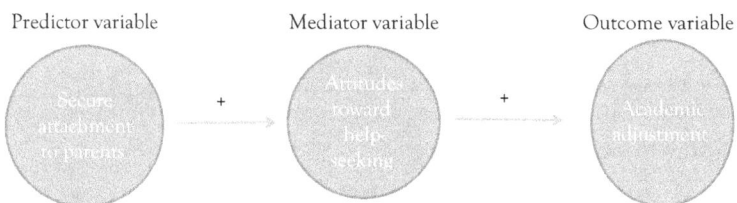

Figure 3.2 An example of a longitudinal study: Examining the relationship between secure attachment to parents, attitudes toward help-seeking, and academic adjustment

Source: Holt (2014).

through helping-seeking attitudes to greater academic adjustment. Holt found evidence to support her model, showing that secure attachment was related to greater help-seeking, which, in turn, predicted better academic adjustment. These results are bolstered by the longitudinal design of the study, providing confidence that attachment predicts help-seeking and adjustment outcomes, rather than the other way around.

Some researchers have been more ambitious and have sought to study students over multiple years of college life. In one exemplar series of publications, Ken Rice and colleagues studied 130 students attending a small, private college from their first to their third year of college (Rice 1992; Rice et al. 1995). Rice was able to successfully track down 81 of the original 130 students in their junior year, a retention rate of 62 percent. It is not uncommon when conducting a longer term longitudinal study that a fair amount of the participants fail to participate in the later follow-up assessments, a phenomenon referred to as *attrition* in methodological nomenclature. It is important that the researcher is able to demonstrate no differential attrition problems in this case, meaning that students who fail to continue in the study are somehow different from those who continue to participate in the study. In his study, Rice found no differential attrition effects. Rice studied secure attachment to and separation–individuation from parents as predictors of academic, social, and personal adjustment. He found first that separation and individuation from parents increased over time, equally for men and women, while secure attachment levels remained stable from first to third year. These results are very consistent with some of the studies I reviewed earlier regarding developmental change in college students, which showed that autonomy and individuation does grow across the years of college, but not at the expense of maintaining a secure relationship with parents (Taub 1995, 1997). Additionally, Rice found that secure attachment to parents in freshmen year predicted greater academic and emotional adjustment two years later (Rice et al. 1995).

A final longitudinal study is remarkable for the extended period of time over which students outcomes were assessed. Gerdes and Mallinckrodt (2001) assessed 208 students regarding their experiences in college during their freshmen year, including contact with faculty, satisfaction with course selection and extracurricular activities on campus, feelings of confidence about being able to handle academic challenges, anxiety levels, and sleep difficulties (college experiences and personal adjustment to college in freshmen year served as predictor variables in this study). Six years later, Gerdes and Mallinckrodt examined the academic records of these students, to determine students' persistence at the university (whether they had completed their degrees [or were still working on them] or had dropped out of the university) as well as the students' academic standing (defined as good-standing versus poor-standing based on whether the student had received a number of failing grades or had been put on academic probation for at least one semester). Gerdes and Mallinckrodt (2001) found that good-standing students persisted at the university for different reasons than poor-standing students. For good-standing students, contact with faculty, satisfaction with course availability, and feeling confident about handling academic challenges predicted greater persistence at the university. For poor-standing students, not sleeping well, feeling anxious, and being dissatisfied with extracurricular activities predicted a lack of persistence at the university. These results suggest that the decision to remain at a particular university has more do with the qualities of the university for students who are doing well academically, whereas it has more to do with the qualities of the student (most particularly their personal adjustment levels) for students who are not doing well academically.

Individual Student Characteristics that Affect the College Adjustment Process

In addition to examining predictor–outcome relationships that occur in cross-sectional and longitudinal studies, college student researchers have examined characteristics of the student that may change the nature of those relationships. In technical scientific language, these individual characteristics are called *moderator* variables, in the sense that these characteristics moderate the relationship between two other sets of variables. As an example, in studying the relationship between secure attachment to parents and depression among college students, a researcher may wonder whether that relationship differs for Caucasians versus students of color or for students who live at home versus residential students. In those cases, ethnicity and residential status serve as moderators of the attachment–depression relationship being observed in that particular study. The examination of moderator variables is extremely important in college student research because it helps qualify the main effects of a predictor variable and provides a more nuanced picture of how adjustment works for particular subgroups of students. In the following subsections, I review some of the most important moderator variables that have been examined in college student adjustment research.

Gender

Gender of student is clearly a variable that can moderate the relationship between many predictor and outcome variables of interest to college student researchers. I have already highlighted many findings in this area, where researchers have shown that women and men navigate the process of separation and individuation from parents and formation of autonomy differently and those differences have implications for male and female students' college adjustment.

Commuter Versus Residential Students

When students first attend college, they generally have a choice about where to live. Although many universities strongly encourage first year students to live on-campus (researchers call these students "residential") a significant number of students choose to live at home with their parents during their first year of school and perhaps thereafter (researchers call these students "commuters"). Researchers have suggested that because commuters are not "leaving home," they may experience the college adjustment process differently, perhaps being less lonely and socially anxious during their freshmen year (as shown earlier in Larose and Boivin 1998) and perhaps showing a different process of separation and individuation from parents. The research on parent–student relationships with commuter versus residential students is quite mixed, actually, with some studies showing greater conflict in the parent–student relationships of commuter students, whereas others showing higher levels of attachment for commuter versus residential students. In our meta-analysis, we found that attachment to parents more strongly predicted adjustment for residential students than for commuter students, suggesting that the ongoing support of parents is more important to students who move away from their parents to attend college (Mattanah, Lopez, and Govern 2011).

Caucasians Versus Students of Color

Ethnicity of the student is another issue that has been studied extensively by college adjustment researchers. Many studies in this area focus on the unique challenges faced by ethnic minority students (African American, Latino/a, Native American, and Asian American) adjusting to predominantly White colleges and universities, where subtle forms of discrimination and prejudice continues to be an issue. In more recent years, researchers have devoted attention to the adjustment process of particular ethnic groups, not assuming that adjustment will look the same for all ethnic "minorities." Chapter 4 of this book will be devoted entirely to an exploration of these issues.

First-Generation Versus Second-Generation Students

College students whose parents or guardians never earned a baccalaureate degree are referred to as first-generation students in the sense that they are the first generation in their immediate family to attend college. A second-generation student is defined as a student where at least one of his or her parents earned an undergraduate degree (Clauss-Ehlers and Wibrowski 2007). A number of researchers have contended that first-generation students are at a disadvantage when they enter the college environment because they are less familiar with the "culture of college" (Cabrera and Padilla 2004). These students have not had the opportunity of talking over the college experience with a caregiver who has gone through it themselves. Moreover, first-generation students are more often from ethnic minority groups or immigrant families, and demonstrate less academic readiness for college (Clauss-Ehlers and Wibrowski 2007). For these reasons, considerable research has focused on differences in adjustment patterns for first-generation versus second-generation students as well as examining interventions to help prepare first-generation students for college. In one interesting study, Clauss-Ehlers and Wibrowski (2007) examined the effects of a state-funded, summer academic transition program in helping first-generation, largely ethnic minority, high school students prepare for entrance into Rutgers, the State University of New Jersey. These researchers found that the transition program helped students feel a greater sense of support from academic counselors and peers in preparing for the upcoming transition to college and that the program successfully allowed students to attend Rutgers.

Veteran Versus Civilian College Students

According to recent statistics, more veterans are transitioning from military to student life at this time than at any time since World War II, with over half a million military veterans, active-duty personnel, reservists, and National Guardsmen attending college as of 2008 (Whiteman et al. 2013). This large increase in military personnel attending college is due in part to the post-9/11 Veterans Education Assistance Act of 2008, which provides significant financial assistance to military veterans wishing to attend college who served in the Afghanistan and Iraq wars. Active duty military-personnel and veteran students face unique challenges while attending college, a concern that has caught the attention of higher education counselors and administrators who are trying to make college a more welcoming environment for these students. On account of their unique experiences, as well as the fact that veteran students are often older, these students can feel alienated in the college environment, different from their younger, civilian peers, and sometimes at odds with viewpoints of their fellow students and faculty at the university. Researchers are beginning to focus attention on the adjustment process of military students. One recent study examined the levels of emotional support that military versus civilian students report receiving from their peers and the influence of peer support on psychological distress and academic functioning (Whiteman et al. 2013). This study followed 199 military service/veteran students and 181 civilian students over a three-semester period at a number of public universities in one Midwestern state of the United States. Results indicated that military/veteran students were more likely to be older (average age of 29.41 vs. 23.67 for civilian students), married

(34 percent vs. 8 percent), and, as predicted, military students reported receiving lower levels of emotional support from peers at the initial assessment point when compared with civilian students, although levels of emotional support increased at the same rate for both groups of students. Importantly, although peer emotional support predicted less psychological distress for civilian students, it was unrelated to psychological symptoms for the military students. Whiteman and colleagues conclude that peer support is not enough to address the psychological challenges that military students face and that counseling centers needed to become better equipped to address the unique needs of these students.

Summary

Chapter 3 has examined the landscape of scientific research on students' adjustment to college. I have defined the concept of adjustment as focusing on how students manage the transition to college and experience academic, social, and personal success in ways that promote their attachment to the university and likelihood of persisting at that university. I have also examined college student adjustment across a wide range of functional domains that are relevant to the well-being of college students, whether they are making the initial transition to college or have been there for a number of years. These domains include academic motivation and achievement, social integration and social support, absence of symptomatic distress and high-risk behaviors, and the ability to advance developmental goals of emerging adulthood, including forming one's personal identity, separating from parents in a healthy way, and exploring career options. Finally, I have focused some attention on how scientists conduct research in this area, identifying widely used measurement instruments, evaluating the major paradigms for conducting research in this field, and discussing important characteristics that affect the adjustment process and forecast different patterns of adjustment for specific subgroups of students.

One important characteristic that has been mentioned a few times is student ethnicity. We know that students of color do not experience college the same way as Caucasian students do and may experience heightened adjustment challenges associated with prejudice and discrimination inherent at many predominantly White institutions they are attending. On the other hand, as noted in Chapter 1, ethnic minority students are attending college at higher rates than ever before in U.S. history and most navigate the college experience with a great deal of success. Hence, we need to understand the resources they tap into, to be able to overcome whatever additional challenges they face in adjusting to predominantly White colleges and universities. The next and final chapter of this book explores the unique adjustment experiences of students of color at American universities.

CHAPTER 4

College Adjustment for Students of Color

Pauline Minnaar and Jonathan Mattanah

Changing Rates of Students of Color Attending College

Over the course of the last 30 years, we have witnessed a dramatic increase in the number of ethnic minority students attending college in the United States. As mentioned in Chapter 1, during the last three decades, the percentage of Hispanic students attending college has increased by 10 percent, Asian American students 4 percent, and African American students 5 percent. These changes are reflective of some of the broader changes in the ethnic makeup of the U.S. population. In 2012, Asians were the fastest growing ethnic group in the United States with an increase of 2.9 percent from the previous year, while Hispanics grew by 2.2 percent (U.S. Census Bureau 2013). In addition, it is likely that the number of ethnic minority students attending college will continue to increase in the coming years given that, according to a 2012 report, approximately 50.4 percent of children under the age of one and 49.5 percent under the age of five were ethnic minorities (U.S. Census Bureau 2012). As these children grow and progress through their education, it will be increasingly important that higher education professionals and college campuses are equipped with the knowledge and resources to promote their success in college.

While these numbers promise to increase the diversity of college campuses across this country, many ethnic minority students experience high attrition rates and continue to be less likely than their White peers to actually earn college degrees. Reports from multiple time points indicate this consistent trend. For example, a U.S. Department of Education report from 2000 indicated that high school completion rates for African American students aged 25 to 29 years increased by about 30 percent between 1971 and 1999, yet the percentage of African American students in the same age range who completed a four-year college degree only increased by about 5 percent (NCES Condition of Education Report 2000). Similarly, in a sample of students who entered college at four-year institutions in 2001, 42 percent of the African American students completed a degree within six years compared to 60 percent of the White students (NCES Condition of Education Report 2010). Similar trends are also seen in the Hispanic community where only 15 percent of the population hold a bachelor's degree or higher, compared to 41 percent of Whites (NCES Condition of Education Report 2015). This is in spite of the fact that Hispanics represent one of the fastest growing ethnic groups in the United States.

A recent report indicated that college completion rates for African American (40 percent), Hispanic (51 percent), and Native American (40 percent) students were significantly lower than that of White (62 percent) and Asian American (70 percent) students (NCES Condition of Education Report 2014). As such, high school graduation and college admission rates may mask some of the persistent social and historical inequities that plague students of color after they enroll at a college or university. These challenges impact the ability of students of color to complete their degrees alongside their peers.

Outline of the Chapter

In this chapter, we will take a closer look at some of the adjustment challenges that are unique to students of color, as well as some protective factors that serve as assets in enhancing their college experience. The first section of this chapter examines the impact of family dynamics and family background on the college experience for students of color, discussing the ways in which these relationships can both challenge and support adjustment. We will then examine some of the unique experiences that students of color who attend predominantly White institutions (PWIs) face. In this section, we discuss the impact of discrimination and prejudice on campus in both social and academic settings, as well as communities of support on predominantly White campuses. The final section of the chapter examines the decision to attend historically black colleges and universities (HBCUs) and the impact of these campuses on the adjustment outcomes and identity development of students of color.

The Impact of Family Factors on Adjustment for Students of Color

We know that the particular family dynamics of students of color, where interdependence and connectedness tend to be highly emphasized, exert an important influence on the college adjustment process. In the following section, we discuss ways in which these dynamics add both stress and support for students of color transitioning to college.

Maintaining Family Connections

One way in which families of color differ from White families is that family structure tends to be more collectivistic with a focus on support, cohesion, interdependence, respect, and intimacy (Diaz, Lizardim, and Rivera 2008; Love 2008; Love et al. 2009; Markus and Kitayama 1991). This difference in family structure continues to affect individuals even when they leave home to attend college. For example, students of color may be expected to maintain a close connection and commitment to their family even if they are no longer living at home (Brower and Ketterhagen 2004; Guiffrida and Douthit 2010; Jackson, Smith, and Hill 2003). Family members of students of color expect the student to remain just as involved in family issues as they were while they were still living at home. For some students, this expectation results in numerous and extended phone calls to their family members and, for others, it can involve frequent visits back home or visits to campus from their family.

These expectations can make it difficult for the student to make the time commitment and emotional investment that being in a new environment demands (Guiffrida and Douthit 2010; Jackson, Smith, and Hill 2003). If students are unable to effectively integrate themselves into the social and academic demands of their college, they may struggle ultimately to be successful in this environment. Furthermore, these conflicts between home and college life can be compounded when students of color are also first-generation or immigrant students. It may be challenging for both the student and their family members to become familiar with balancing these demands without any prior exposure to the world of higher education in the United States (Dennis, Phinney, and Chuateco 2005; Kenny and Perez 1996). It should be kept in mind that the values, priorities, and expectations of the world of higher education have been shaped to a large extent by White people, making the task of integration that much more challenging for first-generation students of color and their families.

Families as Assets: Support and Protection

While the expectation to maintain consistent communication with family members may present challenges associated with balancing college and home responsibilities, it is important to keep in mind that families are also important sources of support. Families serve as important assets, particularly for students who come from more interdependently oriented racial and ethnic backgrounds where close family ties are the norm (Kenny and Perez 1996; Love et al. 2009).

A qualitative study examined ways in which families influence Black college students' academic achievement and persistence in college (Guiffrida 2005). Students who were high academic achievers consistently discussed receiving emotional, financial, and academic support from their families, regardless of their income or levels of education. Further, these students noted that these forms of support were some of their most significant assets as they progressed through college. Family members who encouraged students to focus on their studies and to view their academic success as their most important obligation to both their family and the larger Black community contributed to these students' ability to become high achievers (Guiffrida 2005). Conversely, students in this study who were low academic achievers frequently reported the lack of support from their families as a contributing factor to their poor performance and eventual attrition. Furthermore, some of the low achievers attributed their poor academic performance to long hours that they were working at on- and off-campus jobs in order to pay for their tuition or fulfill an obligation to provide financial support to their family members at home. In this way, the students felt that their educational efforts were not valued, and lacking the emotional and financial support from their family, were at risk for academic underachievement and ultimately a lack of persistence at the university (Guiffrida 2005).

This study illustrates some of the complex ways in which family dynamics interact with college students' academic success and it demonstrates the importance of family support while at college. It also highlights some of the potentially negative effects that family pressures and expectations can have on students' adjustment. Similarly, other researchers have also noted that family pressures to achieve or fulfill financial responsibilities can make it challenging for students to effectively navigate the transition process (Kenny and Perez 1996) and can ultimately cause the student to develop feelings of disloyalty if they do choose to fully explore the college culture (Hinderlie and Kenny 2002). It is worth remembering that for students of color, family support is likely to include a broader collective network beyond immediate family members. An expansive support system that includes community members and extended family may serve as a crucial source of support throughout a student's college transition. It may also act as a source of stress, particularly if the student is primarily motivated to do well in school in order to meet the expectations of others (Markus and Kitayama 1991).

Racial Socialization and College Adjustment

Another way in which families influence the adjustment outcomes of ethnic minority students is through racial socialization, which is the process by which ethnic minority children receive informal education about the values and beliefs surrounding race and ethnicity from their parents (Hughes et al. 2006). Racial socialization involves messages that promote a sense of cultural empowerment and pride, which encourages individuals to develop a healthy racial identity. In terms of social adjustment and academic outcomes (particularly at PWIs), racial identity has been associated with enhanced functioning (Chavous et al. 2002). Racial socialization can also include protective messages, which promote an awareness of societal oppression and can serve as preparation for the possibility of experiencing discrimination as a minority status individual (Anglin and Wade 2007).

The process of racial socialization in Black families, for example, has been shown to positively influence college adjustment by enhancing self-esteem and pride in one's heritage while simultaneously preparing an individual for the reality of encountering discrimination, particularly at PWIs. For example, Anglin and Wade (2007) found that African American students who had received racial socialization messages from their parents or primary caregivers while growing up tended to feel more satisfied with their academic performance and their overall purpose in college than students who had not received such messages. In addition, racial socialization during childhood has also been associated with increased self-esteem and psychological well-being during adolescence and enhanced ethnic group attachment in adulthood (Anglin and Wade 2007; Demo and Hughes 1990).

Furthermore, racial socialization influences the way in which students develop a racial identity, which refers to the extent of identification that an individual has toward their racial group, a process that Black students undergo to a greater extent than their White peers (Van Camp, Barden, and Sloan 2010). Racial identity begins in childhood and families play an important role in guiding students as they start determining who they are both as an individual and as a member of a racial or ethnic group. Entering college with a racial identity and sense of self that is already well developed can enhance a student's ability to thrive at college, particularly if that student is part of an ethnic minority on a majority White campus.

Parental Attachment

Healthy attachment bonds and supportive relationships with parents influence college adjustment for students of color in similar ways that their White peers are influenced. Students with secure attachment bonds tend to adapt more easily to changes in their environment and are more easily able to navigate through milestones such as choosing career fields and formulating adult identities (Berk 1998; Love et al. 2009). In a study that examined parental attachment and psychological distress among African American college students, the results indicated that overprotection and parental invasiveness enhanced psychological distress, whereas parental care and warmth protected against distress (Love 2008). Similarly, maternal overprotection has been shown to predict major depressive disorder in a sample of Latino college students (Diaz, Lizardi, and Rivera 2008). Experiencing supportive parental relationships has implications for African American students' peer relationships as well. For example, one study found that supportive parent–student relationships were mirrored in supportive peer relationships, whereas alienated parent–student relationships were mirrored in alienated peer relationships as well (Love et al. 2009).

In addition to parental attachment in general, research with African American students has found that compared to young adults who were not close to their father figures, those that have a warm and supportive relationship with a father figure reported enhanced adjustment during the transition to college and were twice as likely to enter college or find stable employment after high school (Furstenberg and Harris 1993; Love 2008).

Research focusing specifically on Asian American families has suggested that secure attachment bonds and parental support are associated with lower anxiety and depression (Gloria and Ho 2003; Ying, Lee, and Tsai 2007). Students who have positive emotional connections to their parents are likely to develop similar relationships with their peers as well. In turn, these relationships assist in the development of institutional attachment, which enhances the adjustment experience and social well-being as a whole (Love et al. 2009). Fostering healthy parent–student relationships may be especially important for students who come from both ethnic minority and immigrant backgrounds as it is likely they may experience additional concerns attempting to balance the cultural values of their parents' native culture while also adapting to the value systems of their American peers in the college environment (Ying, Lee, and Tsai 2007).

Unique Experiences of Students of Color Attending Predominantly White Colleges

The overall climate and structure of the institution is another important factor that affects students of color, especially those who attend PWIs. In a recent study examining the adjustment of ethnic minority students to a PWI, Asian American students frequently reported feeling uncomfortable and engaging in fewer social interactions while on campus. Whereas only 11 percent of the Caucasian students stated that they felt shy or uncomfortable on campus, 56 percent of the Asian American participants endorsed this theme at least once (Minnaar 2016). In addition, Asian American students reported significantly greater distress and lower social support and overall college adjustment, indicating lower general well-being and satisfaction on campus than White students.

The reality for many students of color is that, once they arrive on the campus, there are a number of concerns that they must face that their White peers typically do not. Issues related to discrimination and prejudice in the classroom, a lack of representation among faculty and administrative staff, and the pressure to conform to a way of life that differs from their traditional experience while simultaneously managing tasks related to ethnic and self-identity development are all extra hurdles that need to be crossed in order to succeed on campus (Chavous et al. 2002; Kenny and Perez 1996; Love 2008).

Impact of Experiencing Discrimination

In general, the first year of college is filled with academic and social challenges that all students have to face. Basic examples of such challenges include managing greater academic expectations and course loads than what they may have experienced previously in high school, and engaging in the process of forming new relationships and social groups. These broad challenges, however, are especially difficult to manage when they are compounded with direct experiences or observations of racism and prejudice in both academic and social settings. Such events increase the likelihood that students of color will develop perceptions of a hostile campus environment, which can lead to a sense of alienation, a lack of connection or commitment to the institution, and greater levels of psychological, social, and academic difficulties and stress than students who experience more positive interactions on campus (Baber 2012; Hinderlie and Kenny 2002; Pascarella and Terenzini 1980).

The campus-wide racial climate and personal experiences with racism and discrimination affect the extent to which students of color feel that they have access to institutional resources and support to assist them throughout their years in college (Bentley-Edwards and Chapman-Hilliard 2015). In terms of social and academic adjustment, researchers have made a case for the importance of a positive racial climate and a sense of institutional attachment for students of color who endeavor to persist through their studies at PWIs (Chavous et al. 2004; Hurtado and Carter 1997; Kenny and Perez 1996). However, this type of climate is often not the reality at many PWIs. When students from traditionally marginalized backgrounds attend these campuses, it is not uncommon that they find themselves as "outsiders" on a campus that possesses preferences, values, and a history that (whether intentional or not) often exclude those who are dissimilar from the institution's dominant culture (Baber 2012). These feelings and perceptions can affect both students' self-esteem and their persistence in college and eventual degree attainment.

The Academic Setting

Academic contexts in particular tend to be places in which students of color are aware of their minority status at a PWI. Even when they experience academic success, many students report feeling the need to continuously prove themselves to their White peers and instructors in order to overcome experiences of racism and stereotyping and low academic expectations (Baber 2012; Cerezo and McWhirter 2012). Students of color cite more examples of hostility in academic settings than in social contexts, including faculty making stereotypical comments about an ethnic group or failing to incorporate perspectives of diverse cultures in the curriculum (Baber 2012; Guiffrida and Douthit 2010). These experiences of insensitivity make it challenging to form strong relationships with faculty, which is essential for student success in college, particularly in terms of academic success (Hinderlie and Kenny 2002). At PWIs, where most faculty are White, students of color may perceive their faculty as culturally insensitive and experience a range of negative emotions in reaction to threatening experiences in their classes (Bentley-Edwards and Chapman-Hilliard 2015). Students of color struggle to view their faculty as realistic role models, when they have relatively little access to instructors from diverse ethnic backgrounds, who can act as mentors and

provide guidance regarding what it takes to be successful in higher education as a person of color (Sedlacek 1999 as cited in Hinderlie and Kenny
2002; Guiffrida and Douthit 2010).

Researchers have studied the notion that academic success at college
is influenced by more than just academic ability. Bowen and Bok (1998)
studied the class ranks of Black students at PWIs and found them to be
consistently lower than those of their White peers even after controlling
for variables such as high school GPA, SAT scores, gender, and field of
study. The fact that their academic performance was consistently lower in
spite of controlling for a number of variables indicates that there are challenges beyond academic preparation and aptitude that affect the ability
of students of color to achieve academic success, particularly at institutions where they represent a racial or ethnic minority. For some students
of color, the experience of attending a PWI may present developmental
challenges such as identity conflict, and strong feelings of isolation that
can impede their ability to succeed academically (Chavous et al. 2002).
Academic success can also be adversely affected by stigma consciousness, which refers to the extent to which an individual is self-conscious
about their stigmatized status (Brown and Lee 2005). Researchers studied a group of 128 stigmatized (Hispanic and Black) and nonstigmatized
(White and Asian) students to determine whether academic achievement
was affected by stigma consciousness. Their results indicated that among
the Black and Hispanic students, college GPA was negatively associated
with feelings of stigma consciousness but that the same finding was not
true for their nonstigmatized peers (Brown and Lee 2005). Having an
acute awareness of one's status as a stigmatized or minority individual on
campus and experiencing feelings of isolation associated with this awareness can be mentally and emotionally distressing, which can ultimately
have a negative effect on academic success.

Ethnic Identity Development

We discussed previously the influence of racial socialization and racial
identity development on college adjustment. Ethnic identity is another
construct that may be an especially important predictor of well-being
for students of color because of its impact on one's ability to develop a

sense of cultural pride and protection against discrimination (Molix and Bettencourt 2010). Ethnic identity includes possessing a sense of membership with an ethnic group along with the attitudes and feelings associated with that membership, which are often characterized by a process of exploration and commitment to that identity (Phinney 1992). Researchers have studied ethnic identity development and its impact on college students. In a sample of 229 diverse ethnic minority students, researchers found that students who possessed a less-developed sense of ethnic identity and had fewer relationships with peers, mentors, and their larger community also reported lower levels of physical and psychological well-being (Schmidt et al. 2014). In similar studies, other researchers have also found a relationship between higher levels of ethnic identity and greater overall well-being and self-esteem (Phinney et al. 2001; Smith and Silva 2011).

Specific subtypes of ethnic and racial identities, including a multicultural identity characterized by a sense of connection to other cultural groups, and an Afrocentric identity which focuses specifically on taking an Afrocentric perspective in most aspects of life, have also been examined in the literature (Anglin and Wade 2007; Pope 2000). In a sample of 141 African American students attending a PWI, researchers found that an internalized multicultural racial identity was associated with better overall adjustment to college and that students who endorsed this identity felt more satisfied with their social ties at college. Conversely, an internalized Afrocentric racial identity was associated with poorer college adjustment (Anglin and Wade 2007). These results suggest that on predominantly White campuses, students of color who strongly embrace their ethnic identity or perspective may feel isolated and not supported in their immediate environments. Furthermore, the results suggest that developing a more inclusive racial identity that involves establishing ties to other marginalized cultural groups (i.e., sexual and religious minorities) may enhance adjustment by promoting a sense of belonging and connectedness to the larger campus community.

Communities of Support

In response to potential or real experiences of discrimination, many students of color seek to affiliate themselves with groups where they do not feel as though they are outsiders. In particular, affinity group organizations, where students of color can connect predominantly with members of their own culture and community (Guiffrida and Douthit 2010; Van Camp, Barden, and Sloan 2010) provide important opportunities for students to enhance their personal strengths and seek support systems that can alleviate the distress associated with experiencing oppressive social and academic environments (Schmidt et al. 2014). These experiences can also help to expose students to the heterogeneity and diversity of their own ethnic group (Baber 2012). Unlike their White peers who tend to use informal associations with others as a method of becoming socially integrated on campus, students of color who are underrepresented at PWIs may find that becoming involved in more formal groups or associations, such as student organizations, allow for greater social adjustment (Guiffrida and Douthit 2010; Hinderlie and Kenny 2002; Murguia, Padilla, and Pavel 1991). These organizations can assist in bridging the gap between the students' home environments and that of their new campus.

Research on Latino students, in particular, supports the idea that intervention programs that involve mentoring and peer support serve as important avenues through which students improve their self-esteem and gain knowledge about resources on campus, which in turn enhances their educational achievement (Cerezo and McWhirter 2012; Santos and Reigadas 2002). Similarly, a qualitative study that examined how Black students adjust to PWIs found that students who participated in Black student organizations were able to give back to other Black students who were experiencing similar adjustment concerns, act as advocates for changes on campus, and connect with professional mentors. Students believed that each of these activities were pivotal in helping them feel as though they were successful and contributing members of the Black community. In addition, these organizations provided students with a break from their predominately White environment and allowed them to socialize in ways that were more comfortable and familiar to them without the fear of experiencing prejudice or perpetuating negative stereotypes (Guiffrida 2003). Building strong support systems and communities on campus can assist in the social adjustment of students of color as they are able to connect with other individuals who understand the unique challenges they face on predominantly White campuses. It is important to note, however, that some researchers suggest that over involvement in student organizations that are geared toward specific ethnic groups may in fact lead to some potentially negative consequences such as diverting attention from academics, particularly for low-achieving students (Flemming 1984; Guiffrida 2004). In addition, while these organizations help to connect students to others with similar racial or ethnic backgrounds, they may also hinder social integration by further isolating the group as a whole from the larger student body (Gloria et al. 1999; Pascarella and Terenzini 2005). In general, however, research suggests that at least a moderate involvement in formalized groups can serve to enhance adjustment on predominantly White campuses.

A recent trend in higher education is the implementation of intergroup dialogue programs as a method of addressing issues of discrimination on campus. Intergroup dialogue is a facilitated group experience which seeks to provide individuals with a safe and structured environment in which to explore social identities and the attitudes associated with them (Dessel and Rogge 2008). Although the purpose of intergroup dialogue is not directly related to enhancing the college adjustment process, research on the topic indicates that intergroup dialogue does in fact have a positive impact on the college experience. In a meta-analysis evaluating the outcomes of intergroup dialogue, the authors found that in four separate studies, students who participated in such groups increased their perspective taking, while students in other studies increased their ability to recognize oppression, value new viewpoints, and enhanced their awareness of social inequalities (Dessel and Rogge 2008). In one study, African American, Latino, and Asian American students reported an increase in positive relationships with White students four years later as well as perceptions of greater commonality with them. For students who come from social identities with a history of being isolated or in conflict, intergroup dialogue may provide an opportunity for open conversation about their differences, which can lead to developing a deeper understanding and appreciation for one another. Having a safe space to explore social identities ultimately has a positive impact on students' overall well-being and adjustment.

First-Generation Students

As discussed earlier, the challenges associated with adjusting to college are often intensified when students of color are also first-generation students. Heading into college, these students are at a disadvantage in terms of not possessing basic knowledge about higher education that may have been passed along from family members who have completed the college experience before them. Students who lack basic information about college (such as cost of attendance, the application process, degree plans, and how to deal with academic and social stressors that contribute to attrition) tend to experience more problematic college transitions (Cerezo and McWhirter 2012; Pascarella et al. 2003). In fact, research suggests that first-generation students at community colleges in particular may be more likely to complete fewer credit hours, have lower college grades, and work more hours per week than their peers whose parents have both completed at least a bachelor's degree (Pascarella et al. 2003).

As first-generation students adjust to their college environment, many report having to deal with shifts in language use, social class, and access to resources that influence their interactions with family members (Cerezo and McWhirter 2012). This may be true even for students of color who are not first-generation students. Because the predominant culture of the campus tends to differ from that of their family background, students may experience a sense of discontinuity between their family and college environments (Kenny and Perez 1996). This means that in addition to familiarizing themselves with the expectations that are placed on them by their college or university, they have to deal with tension with family members who have not undergone similar experiences. First-generation students in particular tend to find that peers, rather than family members, are better able to provide academic resources such as study strategies, class notes, and advice regarding which classes to take, as well as social support to assist in overcoming specific challenges that are unique to these students (Dennis, Phinney, and Chuateco 2005). When first-generation students are also students of color, these sources of support are increasingly important in mitigating experiences of discrimination and creating a sense of community to assist in managing feelings of isolation on campus.

Experiences at Historically Black Colleges and Universities

The previous section discussed issues related to experiencing racism and prejudice on campus and the importance of finding avenues or communities of social support in dealing with these issues. When Black students, for example, choose to attend an HBCU how do their college experiences differ from Black students who attend PWIs? In this section, we review some of the decision processes involved in choosing to attend an HBCU versus a PWI and the benefits associated with attending an HBCU for ethnic minority students.

How Black Students Choose Between an HBCU and a PWI?

While the messages that families convey to students regarding racial and ethnic identity clearly affect students' college decision making, the socio-economic environment in which students of color grow up is influential as well. In general, African American students from lower socioeconomic classes are likely to also come from neighborhoods that are predominantly African American. Conversely, African American students from more affluent families often come from neighborhoods that are predominantly Caucasian (Chavous et al. 2002). The ethnic makeup of communities and schools influences the level of intragroup contact that students experience prior to arriving at their college or university campus. These experiences shape decisions that are particularly salient for Black/African American students, such as the decision to attend an HBCU or a PWI. Students who have not experienced extensive contact with their own racial or ethnic group may desire such intragroup contact more strongly than others who have spent more time engaging in environments that help to develop a positive view of their own racial group (Van Camp, Barden, and Sloan 2010). A qualitative study indicated that Black students who came from predominantly White high schools were more likely to consider attending an HBCU than students who came from primarily Black high schools (Freeman 1999 as cited in Van Camp, Barden, and Sloan 2010). Students who grew up with a lack of contact with other Black individuals may desire the experience of being a member of the racial majority on their college campus, a decidedly different experience from their high school or broader community. Students from a predominantly Black high school may still choose to attend an HBCU but their reasoning is less related to a desire for intragroup contact.

Some researchers have found that the similarity between high school and college environments is an important factor when looking at which type of college a student chooses to attend. Students of color whose high school and neighborhood environments were predominantly White may choose to attend a PWI. Interestingly, research suggests that students may be making a wise choice in deciding to attend a school that reflects their high school demographics as students who experience a sense of congruence between their high school environments and that of their new college have been shown to enjoy better academic and social integration, particularly for those attending PWIs (Chavous et al. 2002; Graham, Baker, and Wapner 1984).

Benefits of Attending an HBCU

One of the benefits of attending an HBCU is that students are able to more easily gain access to other Black individuals, including faculty, staff, and students who can assist in feeling more invested in the Black community as a whole (Bentley-Edwards and Chapman-Hilliard 2015; Van Camp, Barden, and Sloan 2010). These individuals can serve as important sources of cultural information as well as role models as students begin their professional development. Such benefits are often the reason why individuals choose to enroll at HBCUs in the first place as students do not want to be a minority on their college campus and are seeking to develop more expansive social networks with individuals who share similar experiences. Students may also choose HBCUs because they are eager to take classes that incorporate the Black perspective into course material (Tobolowsky, Outcalt, and McDonough 2005; Van Camp, Barden, and Sloan 2010).

Some researchers suggest that in comparison to Black students attending HBCUs, Black students who attend PWIs experience less satisfaction with their campus environments and do not perceive a strong sense of institutional fit, which can ultimately contribute to increased attrition rates (Bentley-Edwards and Chapman-Hilliard 2015; Guiffrida and Douthit 2010). Furthermore, researchers also highlight the idea that HBCUs provide students with environments that are more culturally congruent with their backgrounds and hence lend themselves to enhancing the college adjustment process (Love et al. 2009). In a study that examined 206 Black students attending PWIs and HBCUs, the researcher found that both environments fostered awareness of racial identity but that certain attitudes or beliefs were more prevalent at one type of institution versus the other (Cokley 1999). For example, the students at the HBCU campuses possessed more African-centered and nationalist beliefs, whereas the Black students attending the PWIs scored higher on measures of assimilation. In addition, the students at the HBCUs experienced more positive adjustment to their campus environments than their peers at the PWIs (Cokley 1999). In another study, researchers found that Black students at PWIs who possessed strong African-centered beliefs and an overall Afrocentric identity were less well adjusted to their campuses and ultimately struggled to experience a strong sense of acceptance and belonging at their institutions (Anglin and Wade 2007). Together these two studies suggest that HBCUs may provide students with an environment that fosters positive adjustment and the opportunity to develop a racial identity that may be viewed more negatively on a predominantly White campus. HBCUs may also highlight experiences of acceptance and unity among the student body, which can foster social opportunities for Black students that mirror what White students experience at PWIs.

The Cokley (1999) study suggests that in some regards (such as racial identity development), students at HBCUs and PWIs are actually quite similar. In line with this idea, Bentley-Edwards and Chapman-Hilliard (2015) found that students in both contexts reported similar levels of involvement in terms of the number of extracurricular and athletic activities in which they were involved. In addition, on both campuses the researchers found that extracurricular involvement was positively related to racial agency, which involves feeling a sense of empowerment to engage in race-based social and political change. Although HBCUs and PWIs may both offer opportunities to foster similar beliefs, students at HBCUs may find that these opportunities are more readily available and that they do not need to seek out specific organizations or events in order to engage in this type of development.

Do HBCUs Provide a Better "Match" for Black Students?

As mentioned previously, one of the potential benefits of attending a HBCU is that Black students are able to enjoy being in an environment that is culturally congruent with their racial and ethnic backgrounds. As such, it is likely that their expectations regarding how to be successful in college will be in line with what their institution expects of them. Conversely, it is not uncommon for Black students attending PWIs to find that they have to negotiate between two domains, one related to their ethnic background and racial identity, and the other related to having to adjust to the values and demands of their predominantly White college environment (Chavous et al. 2002).

In a study that examined the expectations and perceptions of fit between students and their institutions, Brower and Ketterhagen (2004) found that Black students at HBCUs and White students at PWIs tended to succeed quite easily at their respective campuses. These students appeared to be in tune with the expectations of their environments, such as setting realistic goals for their first semester GPAs and finding a balance between being on their own, spending time with college friends, and managing contact and visits with family. In contrast, the Black students who attended PWIs tended to report struggling to succeed in their environment and having to learn the norms and expectations inherent at their university. These students also tended to spend more time alone working on academic tasks and did not have large social networks (Brower and Ketterhagen 2004). While the White students at the PWIs and the Black students at the HBCUs possessed relatively large networks of friends, the Black students at the PWIs appeared to have smaller but relatively close-knit circles of friends. This study reflects some of what has already been discussed with regards to HBCUs being able to potentially provide students with an adjustment experience that more readily incorporates attitudes and beliefs that are already familiar to them as well as greater opportunities for social engagement. For many students, attending an HBCU fosters feelings of belongingness, social well-being, and institutional attachment, while also offering an environment with fewer instances of discrimination or hostility toward racial background (Love et al. 2009).

Summary and Conclusions

In this chapter, we have examined what the college adjustment process is like for students of color and how this experience differs from that of White students. It is clear from the research that there are a number of concerns that can make the college transition challenging for students from ethnic minority backgrounds who attend PWIs. Some students struggle to balance their academic and social adjustment with the expectations that their family members place on them in terms of maintaining frequent contact with relatives back home or assisting with family finances. Other students struggle to find their place on campus while managing threats of discrimination and a lack of awareness of their cultural identity from faculty, staff, and their peers. For students of color who are also first-generation students, these concerns are compounded by the fact that they may not fully understand what to expect out of the college experience. In spite of these challenges, students of color also show resiliency in their ability to form close-knit relationships with select groups on campus, including student organizations and ethnic minority faculty members. These supportive communities play an important role in assisting students with the college transition process by providing an environment that is more similar to their cultural background than the campus setting as a whole. In addition, these communities help students to develop a positive racial identity, which enhances their overall self-concept and general identity development.

For students who attend HBCUs, being in an environment that actively incorporates attitudes and beliefs that are similar to their cultural background has positive effects on their overall adjustment and sense of belonging on campus. While some students who come from majority White high schools and neighborhoods choose to attend HBCUs to become more exposed to this type of environment, others choose PWIs because they offer a greater sense of congruence between their high school and college environments. Understanding the motivation for either choice is a useful point to consider as it informs the kinds of experiences that students of color may seek out as they progress through their college career. Whether it is at a PWI or HBCU, higher education professionals are encouraged to consider how the campus climate impacts the adjustment and overall well-being of students of color and how outreach opportunities can enhance these outcomes.

References

ACHA (American College Health Association). 2014. *National College Health Assessment-II*. Retrieved from www.acha-ncha.org

Adjust n.d. "In *Merriam-Webster's Online Dictionary*." Retrieved from www.merriam-webster.com/dictionary/adjust

Anderson, N. 2015. Duncan Wants Greater Accountability in Higher Education. Easier said than Done. *The Washington Post*, July 28. Retrieved from www.washingtonpost.com/news/grade-point/wp/2015/07/28/duncan-wants-more-accountability-in-higher-education-easier-said-than-done

Anglin, D.M., and J.C. Wade. 2007. "Racial Socialization, Racial Identity, and Black Students' Adjustment to College." *Cultural Diversity and Ethnic Minority Psychology* 13, no. 3, pp. 207–15.

Arnett, J.J. 2000. "Emerging Adulthood: A Theory of Development from the Late Teens Through the Twenties." *American Psychologist* 55, no. 5, pp. 469–80.

Arnett, J.J. 2015. *Emerging Adulthood: The Winding Road from the Late Teens Through the Twenties* 2nd ed. Oxford: Oxford University Press.

Aspinwall, L., and S. Taylor. 1992. "Modeling Cognitive Adaptation: A Longitudinal Investigation of the Impact of Individual Differences and Coping on College Adjustment and Performance." *Journal of Personality and Social Psychology* 63, no. 6, pp. 989–1003.

Association of American Medical Colleges. 2014. *State of Women in Academic Medicine*. Retrieved from www.aamc.org/publications

Astin, A. 1993. *What Matters in College: Four Critical Years Revisited*. San Francisco, CA: Jossey-Bass.

Astin, A. 1999. "Student Involvement: A Developmental Theory for Higher Education." *Journal of College Student Development* 40, no. 5, pp. 518–29.

Baber, L.D. 2012. "A Qualitative Inquiry on the Multidimensional Racial Development Among First-Year African American college Students Attending a Predominately White Institution." *Journal of Negro Education* 81, no. 1, pp. 67–81.

Baxter Magolda, M. 1992. *Knowing and Reasoning in College: Gender-Related Patterns in Students' Intellectual Development*. San Francisco, CA: Jossey-Bass.

Beck, A., R. Steer, and G. Brown. 1996. *Manual for the Beck Depression Inventory-Second Edition*. San Antonio, TX: The Psychological Corporation.

Belenky, M., B. Clinchy, N. Goldberger, and J. Tarule. 1986. *Women's Way of Knowing*. New York: Basic Books.

Bentley-Edwards, K., and C. Chapman-Hilliard. 2015. "Doing Race in Different Places: Black Racial Cohesion on Black and White College Campuses." *Journal of Diversity in Higher Education* 8, no. 1, pp. 43–60.

Berk, L. 1998. *Development Through the Lifespan*. Boston: Allyn & Bacon.

Blos, P. 1967. "The Second Individuation Process of Adolescence." *Psychoanalytic Study of the Child* 22, pp. 162–86.

Booth, R. 1983. "An Examination of the College GPA, Composite ACT Scores, IQs, and Gender in Relation to Loneliness of College Students." *Psychological Reports* 53, pp. 347–52.

Bowen, W.G., and D.C. Bok. 1998. *The Shape of the River: Long-Term Consequences of Considering Race in College and University Admissions*. Princeton, NJ: Princeton University Press.

Braithwaite, S., R. Delevi, and F. Fincham. 2010. "Romantic Relationships and the Physical and Mental Health of College Students." *Personal Relationships* 17, no. 1, pp. 1–12.

Brandy, J., S. Penckofer, P. Solari-Twadell, and B. Velsor-Friedrich. 2015. "Factors Predictive of Depression in First-Year College Students." *Journal of Psychosocial Nursing and Mental Health Services* 53, no. 2, pp. 38–44.

Brower, A.M., and A. Ketterhagen. 2004. "Is There an Inherent Mismatch Between How Black and White Students Expect to Succeed in College and What Their Colleges Expect from Them?" *Journal of Social Issues* 60, no. 1, pp. 95–116.

Brown, R.P., and M.N. Lee. 2005. "Stigma Consciousness and the Race Gap in College Academic Achievement." *Self and Identity* 4, no. 2, pp. 149–57.

Cabrera, N., and A. Padilla. 2004. "Entering and Succeeding in the 'Culture of College': The Story of Two Mexican Heritage Students." *Hispanic Journal of Behavioral Sciences* 26, no. 2, pp. 152–70.

Carver, C., M. Schreier, and J. Weintraub. 1989. "Assessing Coping Strategies: A Theoretically Based Approach." *Journal of Personality and Social Psychology* 56, no. 2, pp. 267–83.

CCMH (Center for Collegiate Mental Health). 2013. *2013 Annual Report* (Publication No. STA 14–43). Retrieved from ccmh.psu.edu

Cerezo, A., and B.T. McWhirter. 2012. "A Brief Intervention Designed to Improve Social Awareness and Skills to Improve Latino College Student Retention." *College Student Journal* 46, no. 4, pp. 867–79.

Chang, E., J. Lin, E. Fowler, E. Yu, T. Yu, Z. Jilani, E. Kahle, and J. Hirsch. 2015. "Sexual Assault and Depressive Symptoms in College Students: Do Psychological Needs Account for the Relationship?" *Social Work* 60, pp. 211–18.

Chavous, T., D. Rivas, L. Green, and L. Helaire. 2002. "Role of Student Background, Perceptions of Ethnic Fit, and Racial Identification in the

Academic Adjustment of African American Students at a Predominantly White University." *Journal of Black Psychology* 28, no. 3, pp. 234–60.

Chavous, T.M., A. Harris, D. Rivas, L. Helaire, and L. Green. 2004. "Racial Stereotypes and Gender in Context: African Americans at Predominantly Black and Predominantly White Colleges." *Sex Roles* 51, nos. 1–2, pp. 1–16.

Chickering, A., and L. Reisser. 1993. *Education and Identity.* 2nd ed. San Francisco, CA: Jossey-Bass.

Clauss-Ehlers, C., and C. Wibrowski. 2007. "Building Educational Resilience and Social Support: The Effects of the Educational Opportunity Fund Program Among First- and Second-Generation College Students." *Journal of College Student Development* 48, no. 5, pp. 574–84.

Cokley, K. 1999. "Reconceptualizing the Impact of College Racial Composition on African American Students' Racial Identity." *Journal of College Student Development* 40, pp. 235–45.

Cooper, M. 1994. "Motivations for Alcohol Use Among Adolescents: Development and Validation of a Four-Factor Model." *Psychological Assessment* 6, no. 2, pp. 117–28.

Coopersmith, S. 1967. *The Antecedents of Self-Esteem.* Palo Alto, CA: Counseling Psychologists Press.

Cornish, J., M. Riva, M. Henderson, K. Kominars, and S. McIntosh. 2000. "Perceived Distress in University Counseling Center Clients Across a Six-Year Period." *Journal of College Student Development* 41, pp. 104–9.

Cutrona, C. 1989. "Ratings of Social Support by Adolescents and Adult Informants: Degree of Correspondence and Prediction of Depressive Symptoms." *Journal of Personality and Social Psychology* 57, no. 4, pp. 723–30.

Cutrona, C., V. Cole, N. Colangelo, S. Assouline, and D. Russell. 1994. "Perceived Parental Social Support and Academic Achievement: An Attachment Theory Perspective." *Journal of Personality and Social Psychology* 66, no. 2, pp. 369–78.

Demo, D., and M. Hughes. 1990. "Socialization and Racial Identity Among Black Americans." *Social Psychology Quarterly* 53, no. 4, pp. 364–474.

Dennis, J.M., J.S. Phinney, and L.I. Chuateco. 2005. "The Role of Motivation, Parental Support, and Peer Support in the Academic Success of Ethnic Minority First-Generation College Students." *Journal of College Student Development* 46, no. 3, pp. 223–36.

Dessel, A., and M.E. Rogge. 2008. "Evaluation of Intergroup Dialogue: A Review of the Empirical Literature." *Conflict Resolution Quarterly* 26, no. 2, pp. 199–238.

Diaz, N., H. Lizardi, E.C. and Rivera. 2008. "The Relationship Between Parental Bonding and a Lifetime History of Major Depressive Disorder in Latino College Students." *Journal of Ethnic & Cultural Diversity in Social Work: Innovation in Theory, Research & Practice* 17, no. 1, pp. 21–36.

Drum, D., C. Brownson, A. Denmark, and S. Smith. 2009. "New Data on the Nature of Suicidal Crises in College Students: Shifting the Paradigm." *Professional Psychology: Research and Practice* 40, no. 3, pp. 213–22.

Eager, K., J. Lorenzo, S. Hurtado, and M. Case. 2013. *The American Freshmen: National Norms Fall 2013*. Los Angeles: Higher Education Research Institute, UCLA.

Emory University Suicide Statistics. n.d. Retrieved from www.emorycaresforyou. emory.edu/resources/suicidestatistics.html

Ensign, J., A. Scherman, and J. Clark. 1998. "The Relationship of Family Structure and Conflict to Levels of Intimacy and Parental Attachment in College Students." *Adolescence* 33, no. 131, pp. 575–82.

Erdely, S. 2014. "A Rape on Campus: A Brutal Assault and Struggle for Justice at UVA." *Rolling Stone* 1223, pp. 68–77.

Erdur-Baker, O., C. Aberson, J. Barrow, and M. Draper. 2006. "Nature and Severity of College Students' Psychological Concerns: A Comparison of Clinical and Nonclinical National Samples." *Professional Psychology: Research and Practice* 37, no. 3, pp. 317–23.

Erikson, E. 1968. *Identity: Youth and Crisis*. New York: Norton.

Fass, M., and J. Tubman. 2002. "The Influence of Parental and Peer Attachment on College Students' Academic Achievement." *Psychology in the Schools* 39, no. 5, pp. 561–74.

Flemming, J. 1984. *Blacks in College: A Comparative Study of Students' Success in Black and White institutions*. San Francisco, CA: Jossey-Bass.

Foubert, J., M. Nixon, V. Sisson, and A. Barnes. 2005. "A Longitudinal Study of Chickering and Reisser's Vectors: Exploring Gender Differences and Implications for Refining the Theory." *Journal of College Student Development* 46, no. 5, pp. 461–70.

Furstenberg, F.J., and K.M. Harris. 1993. "When Fathers Matter/Why Fathers Matter: The Impact of Paternal Involvement on the Offspring of Adolescent Mothers." In *The Politics of Pregnancy: Adolescent Sexuality and Public Policy*, eds. A. Lawson, D.L. Rhode, A. Lawson, and D.L. Rhode, 189–215. New Haven, CT: Yale University Press.

Garcia, J., C. Reiber, S. Massey, and A. Merriwether. 2012. "Sexual Hookup Culture: A Review." *Review of General Psychology* 16, no. 2, pp. 161–76.

Gerdes, H., and B. Mallinckrodt. 1994. "Emotional, Social, and Academic Adjustment of College Students: A Longitudinal Study of Retention." *Journal of Counseling and Development* 72, no. 3, pp. 281–88.

Gloria, A., and T. Ho. 2003. "Environmental, Social, and Psychological Experiences of Asian American Undergraduates: Examining Issues of Academic Persistence." *Journal of Counseling & Development* 81, no. 1, pp. 93–105.

Gloria, A.M., S.R. Kurpius, K.D. Hamilton, and M.S. Willson. 1999. "African American Students' Persistence at a Predominantly White University: Influence of Social Support, University Comfort, and Self-Beliefs." *Journal of College Student Development* 40, pp. 257–68.

Graham, C., R.W. Baker, and S. Wapner. 1984. "Prior Interracial Experience and Black Student Transition into Predominantly White Colleges." *Journal of Personality and Social Psychology* 47, no. 5, pp. 1146–54.

Greenberger, E., and C. McLaughlin. 1998. "Attachment, Coping, and Explanatory Style in Late adolescence." *Journal of Youth and Adolescence* 27, no. 2, pp. 121–39.

Guiffrida, D.A. 2003. "African American Student Organizations as Agents of Social Integration." *Journal of College Student Development* 44, no. 3, pp. 304–19.

Guiffrida, D.A. 2004. "How Involvement in African American Student Organizations Supports and Hinders Academic Achievement." *NACADA Journal* 24, nos. 1–2, pp. 88–98.

Guiffrida, D.A. 2005. "To Break Away or Strengthen Ties to Home: A Complex Question for African American Students Attending a Predominantly White Institution." *Equity and Excellence in Education* 38, no. 1, pp. 49–60.

Guiffrida, D.A., and K.Z. Douthit. 2010. "The Black Student Experience at Predominantly White Colleges: Implications for School and College Counselors." *Journal of Counseling & Development* 88, no. 3, pp. 311–18.

Halamandaris, K., and K. Power. 1999. "Individual Differences, Social Support, and Coping with Examination Stress: A Study of the Psychosocial and Academic Adjustment of First Year Home Students." *Personal and Individual Differences* 26, no. 4, pp. 665–85.

Hinderlie, H.H., and M. Kenny. 2002. "Attachment, Social Support, and College Adjustment Among Black Students at Predominantly White Universities." *Journal of College Student Development* 43, pp. 327–40.

Hoffman, J. 1984. "Psychological Separation of Late Adolescents from Their Parents." *Journal of Counseling Psychology* 31, no. 2, pp. 170–78.

Holmbeck, G., and M. Wandrei. 1993. "Individual and Relational Predictors of Adjustment in First-Year College Students." *Journal of Counseling Psychology* 40, no. 1, pp. 73–78.

Holt, L. 2014. "Help Seeking and Social Competence Mediate the Parental Attachment–College Student Adjustment Relation." *Personal Relationships* 21, no. 4, pp. 640–54.

Horowitz, L., S. Rosenberg, B. Baer, G. Ureno, and V. Villasenor. 1988. "Inventory of Interpersonal Problems: Psychometric Properties, and Clinical Applications." *Journal of Consulting and Clinical Psychology* 56, no. 6, pp. 885–92.

Hughes, D., J. Rodriguez, E.P. Smith, D.J. Johnson, H.C. Stevenson, and P. Spicer. 2006. "Parents' Ethnic-Racial Socialization Practices: A Review of Research and Directions for Future Study." *Developmental Psychology* 42, no. 5, pp. 747–70.

Hurtado, S., and D.F. Carter. 1997. "Effects of College Transition and Perceptions of the Campus Racial Climate on Latino College Students' Sense of Belonging." *Sociology of Education* 70, pp. 324–45.

Jackson, L., and A. Hood. 1985. "Iowa Developing Autonomy Inventory." In *The Iowa Student Development Inventories*, ed. A.B. Hood, 3/5–3/8. Iowa City: Hitech Press.

Jackson, A., S. Smith, and C. Hill. 2003. "Academic Persistence Among Native American College Students." *Journal of College Student Development* 44, no. 4, pp. 548–65.

Kenny, M.E., and V. Perez. 1996. "Attachment and Psychological Well-Being Among Racially and Ethnically Diverse First-Year College Students." *Journal of College Student Development* 37, pp. 527–35.

Kett, J. 2003. "Reflections on the History of Adolescence in America." *The History of the Family* 8, no. 3, pp. 355–73.

King, P., and K. Kitchener. 2002. "The Reflective Judgment Model: Twenty Years of Research on Epistemic Cognition." In *Personal Epistemology: The Psychology of Beliefs About Knowledge and Knowing*, eds. B. Hofer, and P. Pintrich, 37–62. Hillsdale, NJ: Erlbaum.

Kroger, J. 1985. "Separation-Individuation and Ego Identity Status in New Zealand University Students." *Journal of Youth and Adolescence* 14, no. 2, pp. 317–26.

Kroger, J., and S. Haslett. 1988. "Separation-Individuation and Ego Identity Status in Late Adolescence: A Two-Year Longitudinal Study." *Journal of Youth and Adolescence* 17, no. 1, pp. 59–79.

Kuncel, N., M. Credé, and L. Thomas. 2005. "The Validity of Self-reported Grade Point Averages, Class Ranks, and Test Scores: A Meta-Analysis and Review of the Literature." *Review of Educational Research* 75, no. 1, pp. 63–82.

Larose, S., and M. Boivin. 1998. "Attachment to Parents, Social Support Expectations, and Socioemotional Adjustment During the High School-College Transition." *Journal of Research on Adolescence* 8, no. 1, pp. 1–27.

Larose, S., and A. Bernier. 2001. "Social Support Processes: Mediators of Attachment State of Mind and Adjustment in Late Adolescence." *Attachment and Human Development* 3, no. 1, pp. 96–120.

Larose, S., F. Guay, and M. Boivin. 2002. "Attachment, Social Support, and Loneliness in Young Adulthood: A Test of Two Models." *Personality and Social Psychology Bulletin* 28, no. 5, pp. 684–93.

Larose, S., and R. Roy. 1995. "Test of Reactions and Adaptation in College (TRAC): A New Measure of Learning Propensity for College Students." *Journal of Educational Psychology* 87, no. 2, pp. 293–306.

Leary, M. 1983. "Social Anxiousness: The Construct and Its Measurement." *Journal of Personality Assessment* 47, no. 1, pp. 66–75.

Leary, M., and R. Kowalski. 1993. "The Interaction Anxiousness Scale: Construct and Criterion-Related Validity." *Journal of Personality Assessment* 61, no. 1, pp. 136–46.

Lease, S., and D. Dahlbeck. 2009. "Parental Influences, Career Decision-Making Attributions, and Self-Efficacy: Differences for Men and Women?" *Journal of Career Development* 36, pp. 95–113.

Liable, D., G. Carlo, and S. Roesch. 2004. "Pathways to Self-Esteem in Late Adolescence: The Ole of Parent and Peer Attachment, Empathy, and Social Behaviors." *Journal of Adolescence* 27, no. 6, pp. 703–16.

Locke, B., J. Buzolitz, P.W. Lei, J. Boswell, A. McAleavey, T. Sevig, and J. Dowis. 2011. "Development of the Counseling Center Assessment of Psychological Symptoms-62 (CCAPS-62)." *Journal of Counseling Psychology* 58, no. 1, pp. 97–109.

Lopez, F. 1997. "Student–Professor Relationship Styles, Childhood Attachment Bonds and Current Academic Orientations." *Journal of Social and Personal Relationships* 14, no. 2, pp. 271–82.

Lopez, F., and M. Gover. 1993. "Self-Report Measures of Parent-Adolescent Attachment and Separation-Individuation: A Selective Review." *Journal of Counseling and Development* 71, no. 5, pp. 560–69.

Love, K.M. 2008. "Parental Attachments and Psychological Distress Among African American College Students." *Journal of College Student Development* 49, no. 1 pp. 31–40.

Love, P., and V. Guthrie. 1999. "Perry's Intellectual Scheme." *New Directions for Student Service* 88, pp. 5–15.

Love, K., K. Tyler, D. Thomas, P. Garriott, C. Brown, and C. Roan-Belle. 2009. "Influence of Multiple Attachments on Well-Being: A Model for African Americans Attending Historically Black Colleges and Universities." *Journal of Diversity in Higher Education* 2, no. 1, pp. 35–45.

Mahler, M., F. Pine, and A. Bergman. 1975. *The Psychological Birth of the Human Infant*. New York: Basic Books.

Mallinckrodt, B. 1992. "Childhood Emotional Bonds with Parents, Development of Adult Social Competencies, and Availability of Social Support." *Journal of Counseling Psychology* 39, no. 4, pp. 453–61.

Marcia, J. 1966. "Development and Validation of Ego Identity Status." *Journal of Personality and Social Psychology* 3, no. 5, pp. 551–58.

Markus, H.R., and S. Kitayama. 1991. "Culture and the Self: Implications for Cognition, Emotion, and Motivation." *Psychological Review* 98, no. 2, pp. 224–53.

Mather, P., and R. Winston. 1998. "Autonomy Development of Traditional-Aged Students: Themes and Processes." *Journal of College Student Development* 39, pp. 33–50.

Mattanah, J., J. Ayers, B. Brand, L. Brooks, J. Quimby, and S. McNary. 2010. "A Social Support Intervention to Ease the College Transition: Exploring Main Effects and Moderators." *Journal of College Student Development* 51, no. 1, pp. 93–108.

Mattanah, J., G. Hancock, and B. Brand. 2004. "Parental Attachment, Separation-Individuation, and College Student Adjustment: A Structural Equation Analysis of Mediational Effects." *Journal of Counseling Psychology* 51, no. 2, pp. 213–25.

Mattanah, J., F. Lopez, and J. Govern. 2011. "The Contributions of Parental Attachment Bonds to College Student Development and Adjustment: A Meta-Analytic Review." *Journal of Counseling Psychology* 58, no. 4, pp. 565–96.

McAleavey, A., S. Nordberg, J. Hayes, L. Castonguay, B. Locke, and A. Lockard. 2012. "Clinical Validity of the Counseling Center Assessment of Psychological Symptoms-62 (CCAPS-62): Further Evaluation and Clinical Applications." *Journal of Counseling Psychology* 59, no. 4, pp. 575–90.

McCormick, C., and J. Kennedy. 1994. "Parent-Child Attachment Working Models and Self-Esteem in Adolescence." *Journal of Youth and Adolescence* 23, no. 1, pp. 1–18

McCormick, C., and J. Kennedy. 2000. "Father-Child Separation, Retrospective and Current Views of Attachment Relationship with Father, and Self-Esteem in Late Adolescence." *Psychological Reports* 86, no. 3, pp. 827–34.

McNally, A., T. Palfai, R. Levine, and B. Moore. 2003. "Attachment Dimensions and Drinking-Related Problems Among Young Adults: The Mediational Role of Coping Motives." *Addictive Behaviors* 28, no. 6, pp. 1115–27.

Meyer, T., M. Miller, R. Metzger, and T. Borkovec. 1990. "Development and Validation of the Penn State Worry Questionnaire." *Behavior Research and Therapy* 28, no. 6, pp. 487–95.

Minnaar, P. 2016. *The Influence of Self-Construal and Social Support on the Adjustment Outcomes of Ethnically Diverse American College Students.* Unpublished manuscript, Department of Psychology, Towson University: Towson, MD.

Molix, L., and B.A. Bettencourt. 2010. "Predicting Well-Being Among Ethnic Minorities: Psychological Empowerment and Group Identity." *Journal of Applied Social Psychology* 40, no. 3, pp. 513–33.

Molnar, D., S. Sadava, N. DeCourville, and C. Perrier. 2010. "Attachment, Motivations, and Alcohol: Testing a Dual-Path Model of High-Risk Drinking and Adverse Consequences in Transitional Clinical and Student Samples." *Canadian Journal of Behavioral Sciences* 42, no. 1, pp. 1–13.

Mothersead, P., D. Kivlighan, and T. Wynkoop. 1998. "Attachment, Family Dysfunction, Parental Alcoholism, and Interpersonal Distress in Late Adolescence: A Structural Model." *Journal of Counseling Psychology* 45, no. 2, pp. 196–203.

Murguía, E., R.V. Padilla. and M. Pavel. 1991. "Ethnicity and the Concept of Social Integration in Tinto's Model of Institutional Departure." *Journal of College Student Development* 32, no. 5, pp. 433–39.

NCES (National Center for Education Statistics). 2000. *The Condition of Education 2000* (NCES 2000–062). Retrieved from www.nces.ed.gov

NCES (National Center for Education Statistics). 2010. *The Condition of Education.* US Department of Education: Washington, DC. Retrieved from www.nces.ed.gov

NCES (National Center for Education Statistics). 2012. *Digest of Educational Statistics.* US Department of Education: Washington, DC. Retrieved from www.nces.ed.gov

NCES (National Center for Education Statistics). 2014. *The Condition of Education.* US Department of Education: Washington, DC. Retrieved from www.nces.ed.gov

NCES (National Center for Education Statistics). 2015. *The Condition of Education.* US Department of Education: Washington, DC. Retrieved from www.nces.ed.gov

Neal, D., W. Corbin, and K. Fromme. 2006. "Measurement of Alcohol-Related Consequences Among High School and College Students: Application of Item Response Models to the Rutgers Alcohol Problem Index." *Psychological Assessment* 18, no. 4, pp. 402–14.

Norvilitis, J., and H. Reid. 2012. "Predictors of Academic and Social Success and Psychological Well-Being in College Students." *Educational Research International* 2012, pp. 1–6. Retrieved from www.hindawi.com/journals/edri/2012/272030/

O'Brien, K., S. Friedman, L. Tipton, and S. Linn. 2000. "Attachment, Separation, and Women's Vocational Development: A Longitudinal Analysis." *Journal of Counseling Psychology* 47, no. 3, pp. 301–15.

O'Donovan, A., and B. Hughes. 2007. "Social Support and Loneliness in College Students: Effects on Pulse Pressure Reactivity to Acute Stress." *International Journal of Medical Health* 19, no. 4, pp. 523–28.

Owen, J., F. Fincham. and J. Moore. 2011. "Short-Term Prospective Study of Hooking up Among College Students." *Archives of Sexual Behavior* 40, no. 2, pp. 331–41.

Pascarella, E., and P. Terenzini. 1980. "Predicting Freshman Persistence and Voluntary Dropout Decisions from a Theoretical Model." *The Journal of Higher Education* 51, pp. 60–75.

Pascarella, E., and P. Terenzini. 2005. *How College Affects Students: A Third Decade of Research*. Vol. 2. San Francisco, CA: Jossey-Bass.

Pascarella, E., G. Wolniak, C. Pierson, and P. Terenzini. 2003. "Experiences and Outcomes of First-Generation Students in Community Colleges." *Journal of College Student Development* 44, no. 3, pp. 420–29.

Perry, W. 1970. *Forms of Intellectual and Ethical Development in the College Years: A Scheme*. New York: Holt, Rinehart, & Winston.

Perry, W. 1981. "Cognitive and Ethical Growth: The Making of Meaning." In *The Modern American College,* ed. A. Chickering and Associates, 76–117. San Francisco, CA: Jossey-Bass.

Phinney, J.S. 1992. "The Multigroup Ethnic Identity Measure: A New Scale for Use With Diverse Groups." *Journal of Adolescent Research* 7, no. 2, pp. 156–76.

Phinney, J.S., G. Horenczyk, K. Liebkind, and P. Vedder. 2001. "Ethnic Identity, Immigration, and Well-Being: An Interactional Perspective." *Journal of Social Issues* 57, no. 3, pp. 493–510.

Ponzetti, J. 1990. "Loneliness Among College Students." *Family Relations* 39, pp. 336–40.

Ponzetti, J., and R. Cate. 1981. "Sex Differences in the Relationship Between Loneliness and Academic Performance." *Psychological Reports* 48, no. 3, p. 758.

Pope, R.L. 2000. "The Relationship Between Psychosocial Development and Racial Identity of College Students of Color." *Journal of College Student Development* 41, no. 3, pp. 302–12.

Pritchard, M., G. Wilson, and B. Yamnitz. 2007. "What Predicts Adjustment Among College Students? A Longitudinal Panel Study." *Journal of American College Health* 56, no. 1, pp. 15–21.

Pryor, J.H., S. Hurtado, L. DeAngelo, L.P. Blake, and S. Tran. 2010. *The American Freshmen: National Norms Fall 2010*. Los Angeles: Higher Education Research Institute, UCLA.

Rice, K. 1992. "Separation-Individuation and Adjustment to College: A Longitudinal Study." *Journal of Counseling Psychology* 39, no. 2, pp. 203–13.

Rice, K., D. Fitzgerald, T. Whaley, and C. Gibbs. 1995. "Cross-Sectional and Longitudinal Examination of Attachment, Separation-Individuation, and College Student Adjustment." *Journal of Counseling & Development* 73, no. 4, pp. 463–74.

Rosenberg, M. 1965. *Society and the Adolescent Self-Image*. Princeton, NJ: Princeton University Press.

Russell, D., L. Peplau, and C. Cutrona. 1980. "The Revised UCLA Loneliness Scale: Concurrent and Discriminant Validity Evidence." *Journal of Personality and Social Psychology* 39, no. 6, pp. 472–480.

Ryan, N., V. Solberg, and S. Brown. 1996. "Family Dysfunction, Parental Attachment, and Career Search Self-Efficacy Among Community College Students." *Journal of Counseling Psychology* 43, no. 1, pp. 84–89.

Santos, S.J., and E. Reigadas. 2002. "Latinos in Higher Education: An Evaluation of a University Faculty Mentoring Program." *Journal of Hispanic Higher Education* 1, no. 1, pp. 40–50.

Schmidt, C.K., S. Piontkowski, T.L. Raque-Bogdan, and K.S. Ziemer. 2014. "Relational Health, Ethnic Identity, and Well-Being of College Students of Color: A Strengths-Based Perspective." *The Counseling Psychologist* 42, no. 4, pp. 473–96.

Schultheiss, D., and D. Blustein. 1994. "Role of Adolescent-Parent Relationships in College Student Development and Adjustment." *Journal of Counseling Psychology* 41, no. 2, pp. 248–55.

Scott, D., and A. Church. 2001. "Separation/Attachment Theory and Career Decidedness and Commitment: Effects of Parental Divorce." *Journal of Vocational Behavior* 58, no. 3, pp. 328–47.

Sedlacek, W.E. 1999. "Black Students on White Campuses: 20 Years of Research." *Journal of College Student Development* 40, pp. 538–50.

Smith, T.B., and L. Silva. 2011. "Ethnic Identity and Personal Well-Being of People of Color: A Meta-Analysis." *Journal of Counseling Psychology* 58, no. 1, pp. 42–60.

Solberg, V., G. Good, and D. Nord. 1994. "Career Search Self-Efficacy: Ripe for Applications and Intervention Programming." *Journal of Career Development* 21, no. 1, pp. 63–72.

Straub, C. 1987. "Women's Development of Autonomy and Chickering's Theory." *Journal of College Student Personnel* 27, pp. 216–24.

Taub, D.J. 1995. "Relationship of Selected Factors to Traditional-Age Undergraduate Women's Development of Autonomy." *Journal of College Student Development* 36, pp. 141–51.

Taub, D.J. 1997. "Autonomy and Parental Attachment in Traditional-Age Undergraduate Women." *Journal of College Student Development* 38, pp. 645–53.

Tennant, J., M. Demaray, S. Coyle, and C. Malecki. 2015. "The Dangers of the Web: Cybervictimization, Depression, and Social Support in College Students." *Computers in Human Behavior* 50, pp. 348–57.

Tinto, V. 1993. *Leaving College: Rethinking the Causes and Cures of Student Attrition.* 2nd ed. Chicago: University of Chicago Press.

Tobolowsky, B.F., C.L. Outcalt, and P.M. McDonough. 2005. "The Role of HBCUs in the College Choice Process of African Americans in California." *Journal of Negro Education* 74, pp. 63–75.

Twenge, J. 2013. "The Evidence for Generation Me and Against Generation We." *Emerging Adulthood* 1, no. 1, pp. 11–16.

U.S. Census Bureau. 2012. *Most Children Younger Than Age 1 Are Minorities, Census Bureau Reports*. Retrieved from www.census.gov/newsroom/releases/archives/population/cb12-90.html

U.S. Census Bureau. 2013. *Asians Fastest-Growing Race or Ethnic Group in 2012, Census Bureau Reports*. Retrieved from www.census.gov/newsroom/press-releases/2013/cb13-112.html

Urchino, B. 2004. *Social Support and Physical Health: Understanding the Health Consequences of Relationships*. New Haven, CT: Yale University Press.

Van Camp, D., J. Barden, and L.R. Sloan. 2010. "Predictors of Black Students' Race-Related Reasons for Choosing an HBCU and Intentions to Engage in Racial Identity-Relevant Behaviors." *Journal of Black Psychology* 36, pp. 226–50.

Vivona, J. 2000. "Parental Attachment Styles of Late Adolescents: Qualities of Attachment Relationships and Consequences for Adjustment." *Journal of Counseling Psychology* 47, no. 3, pp. 316–29.

Wahler, H. 1968. "The Physical Symptoms Inventory: Measuring Levels of Somatic Complaining Behavior." *Journal of Clinical Psychology* 24, pp. 207–11.

Walsh, A. 1995. "Paternal Attachment, Drug Use, and Facultative Sexual Strategies." *Social Biology* 42, nos. 1–2, pp. 95–107.

Weinberger, D., and G. Schwartz. 1990. "Distress and Restraint as Superordinate Dimensions of Self-Reported Adjustment: A Typological Perspective." *Journal of Personality* 58, no. 2, pp. 381–416.

Whisman, M., and E. Richardson. 2015. "Normative Data on the Beck Depression Inventory—Second Edition (BDI-II) in College Students." *Journal of Clinical Psychology* 71, no. 9, pp. 898–907.

White, H., and E. Labouvie. 1989. "Towards the Assessment of Adolescent problem Drinking." *Journal of Studies on Alcohol* 50, no. 1, pp. 30–37.

Whiteman, S., A. Barry, D. Mroczek, and S. Wadsworth. 2013. "The Development and Implications of Peer Emotional Support for Student Service Members/Veterans and Civilian College Students." *Journal of Counseling Psychology* 60, no. 2, pp. 265–78.

Winston, R. 1990. "The Student Developmental Task and Lifestyle Inventory: An Approach to Measuring Students' Psychosocial Development." *Journal of College Student Development* 31, pp. 108–20.

Winston, R., T. Miller, and J. Prince. 1987. *Student Developmental Task and Lifestyle Inventory.* Athens, GA. Student Development Associates.

Ying, Y., P.A. Lee, and J.L. Tsai. 2007. "Predictors of Depressive Symptoms in Chinese- American College Students: Parent and Peer Attachment, College Challenges and Sense of Coherence." *American Journal of Orthopsychiatry* 77, no. 2, pp. 316–23.

Zhong, J., and J. Arnett. 2014. "Conceptions of Adulthood Among Migrant Women Workers in China." *International Journal of Behavioral Development* 38, pp. 255–65.

Index

TITLES FROM OUR PSYCHOLOGY COLLECTION

Children's Rights: Towards Social Justice
by Anne B. Smith

The Elements of Mental Tests, Second Edition
by John D. Mayer

Momentum Press is one of the leading book publishers in the field of engineering, mathematics, health, and applied sciences. Momentum Press offers over 30 collections, including Aerospace, Biomedical, Civil, Environmental, Nanomaterials, Geotechnical, and many others.

Momentum Press is actively seeking collection editors as well as authors. For more information about becoming an MP author or collection editor, please visit http://www.momentumpress.net/contact

Announcing Digital Content Crafted by Librarians

www.ingramcontent.com/pod-product-compliance
Lightning Source LLC
LaVergne TN
LVHW020356230326
834520LV00015B/1273